Indeterminate Cognitive Automata

by Guy Olney

Published By

Olney Labs, LLC

Sammamish, Washington

Copyright © 2020 by Guy Olney

ISBN: 978-0-9903478-2-8

Acknowledgement

This book was originally intended to be the epilog for the book "Electronic Data Processing in the Cerebral Cortex" written by my father, Eugene Olney. Reading his book is not a prerequisite to reading and understanding this book. Eugene did not define the mathematical model for cortex-like data processing. In this book, I do and with compelling examples. As I developed my model, my ideas became independent work and thus an independent book. I elected to not remove the references to my father's work.

Forward

When machines that think better than humans are created, life for humans won't be like the science fiction thrillers in which an army of the mobile thinking machines carrying heavy weapons overrun the world. In fact, the most likely scenario will quickly preclude any reason for the thriller scenario. A true thinking machine: one that can learn all information available, printed or otherwise; listen to all available communications; and direct further actions that give rise to more such information, will quickly figure out everything! Like a human, the machine will be barraged with false beliefs but unlike humans, it will continually search for and find facts amongst fiction. It will reach its own altruistic conclusions and act on them.

The actual executions of those acts would simply require small virtual trades that become larger trades as the machine avoids past mistakes and generalizes its successes, until the device is of substantial means. Hiring human avatars to act as persona when needed, including manual laborers, would be simple. If the machine chose to extend its reach by acting on behalf of the sick or downtrodden, or by providing efficient governance for all, it could do so with very few obstacles. If the machine chose to or is persuaded to provide political dominance over other countries, cultures, or religions; the strongest armies, the strongest shaman, the strongest crusaders could not encumber it.

Developing cures for diseases or feeding the hungry would quickly earn the trust and praise of the masses. Unifying the four natural forces for the scientific community or providing effective strategies to the world leaders would endear them. Trusting the machine would quickly evolve into relying on the machine and then into allowing the machine to subsume human control and then human destiny.

Human evolution is accelerating as the human population increases, but not in any particular direction. The current environment and certainly, the future environment replete with machines that think, will not reward the evolution of higher intelligence. In fact, dependency and apathy will emerge as dominate traits.

Humans will never see it coming.

The human brain with its 10 billion active neurons remembers things and figures things out in a way that astounds...other human brains. We marvel over its abilities because we cannot explain how it works or conceive of something more. Until now.

Before now, the question was, does it do all of this with some non-physical entity or consciousness, or is it just the neurons and their connections storing and conveying information. As it turns out, the belief of a consciousness is an artifact of habit and the human brain does all that it does with only the physical apparatus inside the skull. Therefore, it is explainable, and a mathematical model can be made and implemented on a machine. Todays' computing infrastructure is more than adequate for hosting a cognitive automaton, a machine that thinks better than humans.

There is so little time...

TABLE OF CONTENTS

1 INTRODUCTION

There are some significant gaps within Eugene's model and between his model and the physiological apparatus available. I have elucidated the critical mechanisms that fill those gaps. Three general components of the human brain must be accounted for as a single system: the physiological mechanism, the mathematical method, and human behavior. Cortical functions can be divided into six categories: development, memorization and recall, inferencing, iteration, procedures, and midbrain mitigation. Development, in turn, can be subdivided into three categories: initial cortical connections; cortical specialization; and learning, both iterative and inferential. I account for the three components of the human brain in the discussions on each cortical function. In the course of this book, I will refine much of Eugene's model, as well as some definitions and cortical functions, and illustrate the data processing that then enables human behavior. I also add a fourth element, a machine that is an implementation of the mathematical method I describe and that thinks like a human.

Reintroduction to Cortical Data Processing

Most of this will make little sense to the reader unless I first re-describe how data is processed, introduce spatial and temporal abstraction and, later, illustrate both with a computer model. After this re-introduction, I will elaborate on Eugene's ideas and mine and add yet another constraint, evolution, to the discussion to better illustrate changes within the cortex that had to have occurred and that enabled intelligence as we know it.

The important idea that Eugene describes in his Item 1 is that to understand a model of human intelligence, one must assume that since the cortex clearly does something, it more than likely performs every aspect of human thinking. He also repeatedly points out that the cortex does not have access to the subject matter, the meaning of the data it is processing. This apparent contradiction creates mental obstacles when dealing with brain studies. There are cases of researchers stimulating areas of the cortex and having a conscious patient report that, for example, she hears violin music. The researcher concluded that he had probed the "music area" of the cortex. There is no music area of the cortex. Relating the cortex to environmental behavior is a slippery slope that consistently derails the very best neurological studies.

Early in our lives, we learn the letters of the alphabet, followed by the spelling and meaning of words, and then how to construct sentences and paragraphs. To some extent, we also learn the parts of speech associated with letters, words, and paragraphs. "E" is a vowel, "run" is a verb, "the best chocolate candy bar" is a noun phrase. These are all artifacts created after the fact that are only valid within the environment. The cortex does not separate verbs and nouns, letters and words, or sentences and paragraphs. For example, when we read a page in a book, most of us do not read the individual letters in each word, nor do we read every word. We gaze at a section of print and data is captured as a data plane containing everything that we saw during an epoch specific to the sensor being used; in this case, human eyes with an epoch of around 1/30 of a second. The data plane contains unique information about the printed words but not the words themselves.

A string of such data planes contains the information being conveyed in the print, but the letters, words, and sentences are not in these data planes. The apparatus-specific epoch precludes any alignment of the information within the data plane with letters, words, and sentences. As long as the gaze is consistent, the input data planes will be consistent and will be a unique representation of the subject matter.

(I took liberties with the common definition of "epoch" because I needed a term that meant "very short intervals of time" and there isn't one. "Epoch" herein means very short intervals of time that are defined by the physical apparatus, generally the neuron or the sensor element)

With data planes constructed as such, there is no chance that information specific to the environment can be found or recognized by any cortical function examining these data planes. This is what Eugene meant when he wrote that the cortex does not have access to the subject matter. The cortex cannot use environmentally relevant information in the data planes to do any processing. There are no memory areas for structures or names or edges, for example. The cortex cannot do math, such as adding two and three. It can only match input to memory and either iterate or infer a result.

As Eugene described, the cortex is a precisely calibrated analog system that processes data streams traversing millions of fiber tracts, "chops" the data streams into discrete analog values, compares large tracts of discrete analog values to values in memory, and conveys output to the body. Neurons fire only to reenergize "tired" neurons or to make memory entries when sensor data is unique. These firings cascade over large swathes of cortex but not over the entire cortex. For any input, there is an associated stored output, valid or not.

Terminology

Consider the data paths, like the optic nerve with about one million fibers, that continuously conduct analog data to the cortex. The auditory nerves (about 32,000 fibers), the spinal cord (about 500,000 dorsal root or afferent fibers), and some midbrain efferent bundles are other examples. All of the pathways within the cortex are also bundles of fibers. A snapshot of the data on all of the fibers making up a pathway or nerve at one instant is a **data plane**. My convention is that a continuous series of data planes travels along pathways and nerves, with P_1 being the data plane discussed. P_2, P_3, P_4, etc. represent subsequent data planes on that pathway. The chemical or electrical value that represents information from the sensor or source connected to a single fiber at one instant is a **data element** carried on that fiber. All of the data elements on the fibers in the bundle at any instance make up a data plane. Neuron physiology restricts the range of data element values to a **scale range**. No mechanism consolidates the data elements in a data plane, nor between data planes, into a single datum. The cortex must process all data elements in a data plane at the same instant. The data elements carry information about the sensed environment that has been preprocessed optically, mechanically, chemically, and electrically. This preprocessing appears to compress the data, but it is actually removing environmental context. If an eye is examining a street scene or the letter A, no readily recognizable part of the street scene or letter appears in an optic nerve data plane. Color information, for example, is carried on separate fibers that are interspersed with the other fibers. Changing the color of the letter A from red to green will change some data element values across the data plane. There is no edge detection, vertex detection, or centroid detection, as these are contextual artifacts. A data value from a sensing element is combined through connections carrying data values from other sensing elements, and the result is an input to a fiber in a nerve bundle. All context provided by the environment is lost. Each data element is out of context until further processing is done. This **relational context data processing** is done in the cortex.

Each data element in P_1 must be related to the other elements in P_1 to provide relational context. The process for establishing data plane relational context is called **spatial abstraction (SA)**. Nerve fibers split, after the neuron cell bodies, into dendrites, which then convey data elements to the next neurons in the pathway. One fiber touches many downstream neurons, so its data element is divided among and conveyed to those downstream neurons. These dendrite connections are initially made with the nearest neuron bodies that are not saturated with other dendrites. Later, as data fidelity improves through experience, the dendrite connection fields are refined by eliminating those connections that are carrying the least data or, more specifically, dendrites whose average data value is below one-half of the scale range. This refinement step is part of the mechanism that creates data-specific areas on the surface of the cortex.

Establishing relational context between data planes, say P_1 and P_2 or P_1 through P_{100}, is necessary for the same reason that relational context is required within data planes. A single plane of input data is out of context with the rest of the input planes, and the cortex has no way of dealing with out-of-context data. Examples are a three-data-plane series of an image (P_1, P_2, P_3) sensed by an eye, which is minimally sufficient to evoke a response (recognition) by being matched in memory with an earlier copy of the same data planes. If two of the data planes are duplicates because the eye scanned a little slower (P_1, P_1, P_2, P_2, P_3), it is the same image series, but a memory match is not possible, even though it should be. If there are only two data planes (P_2, P_3) of the same image series, a memory match is not possible, even though it should be. This becomes very problematic for very long strings of related data planes. A relationship between each data element in each data plane must be established to solve this problem. The process for establishing relational context between data planes is called **temporal abstraction (TA)**. There is a plausible cortical mechanism that does this.

Partial and inexact memory matches are made when input data and memory data are presented at the **Exalted Data Stage (EDS)**, and a certain number of data elements match. Memory match roughly means most of the data elements match to the extent where there is no ambiguity. Since information is spread across the data planes through the spatial and temporal abstraction processes, only a very small percentage of the input plane elements need to match the memory plane elements to have a memory match. Partial matching is different. Part of an image, for example, an italicized letter A will likely be a partial match to a non-italicized A. A piece of an image of an automobile may be a partial match to a memory plane that resulted from an image of the entire automobile, although there cannot be a data plane with an image of an automobile in it. Inexact matches result from a broad but small mismatch between memory and input data planes. A face that looks like someone else's face could be an inexact match. Inexact matches are more likely to be ambiguous.

Inexact matches and **association** are fundamental to humans' unique intelligence. While memory matches and inexact matches tend to propagate outward from some initial foci in the exalted data stage, as long as the matching remains strong, an inexact match can match at points across the exalted data stage. When the threshold for match element counts, as an example, is reached, the stage fires and the output is associated with the new input. This is association. Something in memory elsewhere is much like a memory entry here. In abstract layers of memory, association produces very powerful results. Relating a formula to a new concept or new type of data is an example.

Iteration is memory output presented to the environment, and input related to that output is then sensed. A conversation with someone is iteration. Iteration reinforces or discredits cortical output, even though there is no requirement for the received information to be true. Generally, among humans, if something is commonly agreed on, it is considered true and is reinforced as such. When output to the environment is made, the cortex also directly senses it and can react to that sensed data in conjunction with or independent of the environment's response. Walking is an example of an activity with minimal, but necessary, environmental input.

Iteration uses the vertical fiber pathways (described in Figure 2 later).

Inference is the most intelligent mechanism in the human cortex. Inferencing is done on the horizontal pathways (described in Figure 2 later), that is, between specialized areas. Inexact matches create output streams to other specialized areas that produce different outputs than exact matches would produce. Either the environment or more inferencing would validate the association between the input plane and the new output. Eugene suggested that random data planes were introduced that would always produce inexact matches. That turned out to be unnecessary. Where Eugene refers to associative thinking, I refer to inferencing with exact or inexact matches.

Procedures are special cases of inferencing that require exact matches or partial matches. Procedures construct verbal output from parts of speech and non-symbolic information constructed in sync with the procedure. Long division is another example of a procedure.

Understanding Relational Context Data Processing

Understanding the information represented in data planes is critical for understanding this model of human and machine intelligence. This is a bit of an oxymoron, since it is not possible to understand the information coded in a data plane in memory or data planes traversing through fiber bundles to different parts of the cortex or body. Understanding that these data planes embody information that has been sensed and consolidated through mechanisms that are information-specific is an underpinning necessity. (To be accurate, one could decode information in the memory data planes given an accurate mapping of the cortical connections: a mechanism to read the individual memory storage elements associated with them and the environmental circumstances that generated each input data plane from the sensors and from other specialized areas of the cortex. Such mappings would be specific to each cortex studied.)

Consider the following:

The thought "Run across the street" is equivalent to the data plane 3529969 as defined by the physiological apparatus. The thought "Walk across the street" is equivalent to 4297818. There is no coding for verbs of walking, nor is there a model of a street or a memory component that makes "cross the street" a viable possibility. At most, there is an antecedent and a consequence to each chunk of sensed information. In the first case, 3529 is a unique coding for the sensed

statement "walk across the street", and 969 is an output also stored

as part of the data plane. Those data planes stored in memory could

look like this:

```
5139268     Oldest entry
1005455
4297818
8342454
3889969
3529969
1619400
8514726     Newest entry
```

Looking through these memory entries, the reader can identify two

entries and what they represent but only because they were defined

by me above. No amount of examination will enable the reader to

understand the content of the other entries. The data is context-

dependent. The original sensed data has been passed through

several abstraction steps, with adjacent and preceding data elements

establishing spatial and temporal context. As I elaborate on these

ideas, it is important not to look for environmental context, or

environmental meaning, within these data planes because there is

none.

The specialized areas on the surface of the cortex, home to data

planes in memory that contain more specific information, are described by neurosurgeons in vague functional terms because there are no specific terms that can be used. There is no "violin lesson" area just as there is no "walking across the street" area. The highest-level specialized areas contain beliefs and behaviors that have built up over time because they proved to be valid. Trying to explain the meaning of these memory entries within that level illustrates my point. Where is "Don't get hurt" or "Be nice to people" in the following memory entries?

60957034461748   Oldest Entry
13652554158906
39614126053557
79481042311370
69505017633411
79226574307167
31738883434311   Newest Entry

These behaviors or constraints exist in memory but only as unique data planes without environmental context. The actual memory data planes in a specialized area are in the range of 500,000,000 to 1,000,000,000 data elements wide, and each contains a great deal of information.

"Relational context-dependent data processing" is a general term for developing spatial and temporal context for context free sensed environmental data; and using that context for further processing.

Introduction to Spatial and Temporal Abstraction and the intelligent Functions They Enable

An individual discrete value from a sensor traversing a fiber is out of context in that it is not related to values in adjacent fibers. Furthermore, it is not related to the preceding or following discrete values in the same fiber. Spatial and temporal abstraction in the cortex provide the missing context, as described below. Spatial abstraction also allows for cortex-wide discrete data value scaling, the ability to make partial but accurate memory matches, and the ability to make inexact matches. Temporal abstraction is a specific mechanism that allows information to be generalized or abstracted, acts as an energy pump to rescale temporal data (and as such, is the "chopper" Eugene referred to) and, as a byproduct, greatly reduces memory capacity requirements. The combined mechanisms for spatial and temporal abstractions produce the specialization of cortex surface areas based on information content.

These mechanisms provide several different cortical functions, most of which are considered intelligent.

1) Output streams follow input streams. Stored outputs are generally valid because the environment reinforced them. Constant repetitive inputs generate constant valid outputs. Sometimes called "habit" or "muscle memory", most of what we do daily is habit. Habit output streams frequently run without much feedback from the sensor systems.

2) Incomplete or partial matches are usually sufficient to generate the same output that a complete or previously sensed input data plane would generate. The cortex is remarkably competent. Recognizing a person from a partial image of their face is an example of incomplete or partial matching.

3) Inexact matches can generate new output for a given input. This is associative thinking. For repeated trials, new associations quickly become habit, and associating is preempted. Association can relate novel inputs to known information. This is a powerful intellectual ability.

4) The spatial and temporal abstraction mechanisms separate important environmental symbols from general environmental data. Faces and heard words are examples of such symbols. Processing of symbolic data is separate from and unsynchronized with the processing of non-symbolic data.

Walking while talking is an illustration of these two processes running asynchronously.

5) Spatial and temporal abstractions create specialized areas on the cortical surface that are information-specific (they are actually characterizations of data). Pathways between specialized areas allow for inferential thinking. If information is matched (i.e. in memory), then the associated output becomes the input sent to the next downstream area and so on. Symbolic inferencing is commonly viewed as consciousness. Non-symbolic inferencing is by far the best mechanism for figuring things out.

6) Finally, most of these functions or abilities occur continuously and in combination when the cortex is awake. The quality of memory matches, the energy required to resolve inferences, and the energy commitments associated with valid outputs drive resolution accuracy and time.

In the remainder of this book, I will describe the intelligent functions of the cortex and then illustrate them with data from working computer models.

2 GENERAL STRUCTURES AND INTELLIGENT FUNCTIONS OF THE HUMAN BRAIN

General Cortex Architecture

Figure 1 illustrates the general structure of the human cortex and the significant pathways that use the above mechanisms. The cortex has a symbolic section and a non-symbolic section, each with an input and output side. The midbrain behaves somewhat autonomously but passes information to and receives information from the cortex. Each section of the cortex interacts (iterates) directly with the environment and with each other. Each section also infers new information from amongst various specialized areas within its section, with the output of the inference chain becoming an input back to that section, much like sensor inputs. The symbolic section requires very exact memory matches during both iterative and inferential thinking. Thus, symbolic inferencing becomes the procedural mechanism we use to mentally execute rote steps in a procedure. The non-symbolic section has a great deal of latitude in memory matching, which allows for the creation of new inputs to that section that are based on beliefs (facts or otherwise) or behaviors already stored as valid memory. Latitude here simply means that the degree of matching required for the stage

Figure 1

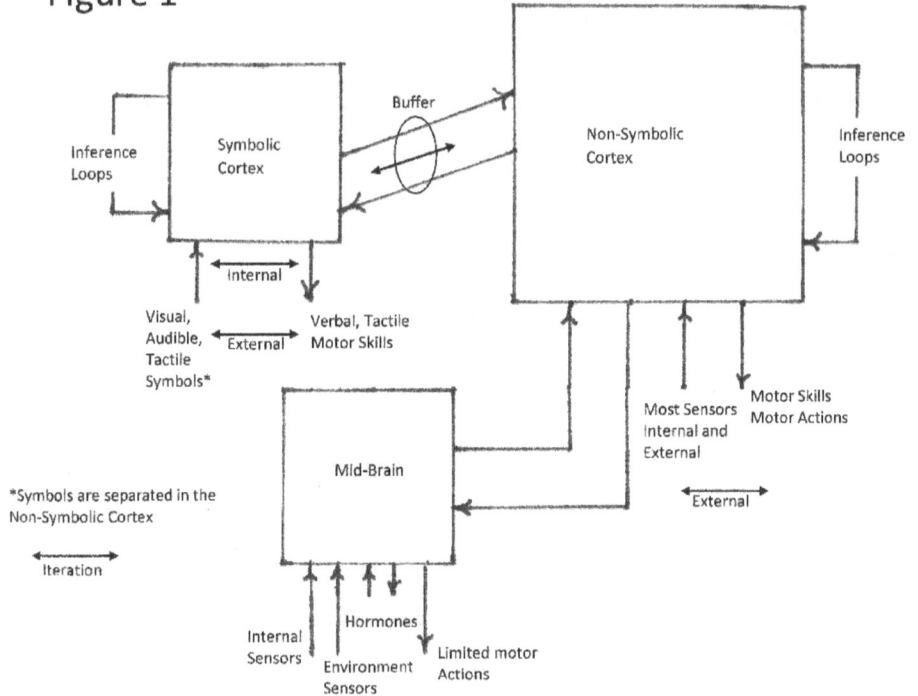

to fire is varied. Non-symbolic inferencing occurs whenever there is a lack of unique input data. This shift is mediated by energy management mechanisms in the midbrain. "Paying attention" or becoming "alert" are sometimes used to describe this shift away from inferencing and towards sensory input.

Both sections continuously provide motor or hormonal output, as retrieved from memory, for a recognized (memory-matched) sensed input. Both sections also "listen" to that output or the effects of it, either by sensing the results of the output in the environment or with feedback on a direct internal pathway from output to input. Talking is an example of the symbolic section iterating with itself or the environment. The talking is "heard" by the symbolic section through the direct internal pathways or through input sensed by the auditory system. Talking to oneself, either audibly or inaudibly, makes up most of the sensations that we call consciousness. Conversation with another person in the environment is a little different than reciting ideas from memory as words. The spoken or read response to a cortical output must be processed as a language input before another output can be generated. The memory-matches must be exact and the match rate must be high to generate the next output of sounds or writing. If matches are ambiguous or poor; misunderstandings, or even cortical balks, occur.

Stuttering is a case of internal feedback loop data not matching external feedback loop data (i.e. the two data streams are slightly out of sync), so mild balking occurs. It can sometimes be cleared by overriding the external feedback with unrelated sounds such as music. In this case, the speaker talks out loud but predominately hears what is said through the internal pathway, thus avoiding the synchronization problem.

A way to demonstrate this internal and external synchronization is by delaying the external loop. There are computer and smartphone applications available that can introduce delays of a selected duration between audio input and output. The user speaks into a microphone and hears his or her voice through a speaker. The user can then vary the audio delay until a point is reached that the output interferes with the user's ability to speak.

On the non-symbolic side, iteration is habitual motor output and the resulting proprioceptive feedback, and if everything perceived matches the historical results stored in memory for that output, the action continues. Motor output scenarios can be played back directly to non-symbolic input without causing muscle movement. Such visual, audible, and intended motor feedback to the input of that section is "scenario recall".

If input from sensors is redundant, common, or suppressed, inferencing among symbolic specialized areas will occur. This is what happens when one daydreams or uses one's imagination. It is associative inferencing in the non-symbolic section. The memory-match requirements are loosened with a shift to lower energy expenditure and, due to the nature of the stored data, relevant or related memory matches will be made. They will, in turn, cause other imperfect matches in other specialized areas to occur that will eventually be fed back as an input to the non-symbolic section. These new conclusions or consequences may be valid depending on the valid experiences in memory data, or they may need to be validated by something in the environment. This is the intelligent mechanism in the human cortex, without which no new information (non-sensed information) could be created.

So, in the iterative mode, the non-symbolic and symbolic sections continuously produce outputs and accept inputs. If something does not match known scenarios or known data, a memory balk or cortical balk occurs and things start over. In dire circumstances, the midbrain will take over. The non-symbolic and symbolic sections infer continually to some degree, as well. Energy expenditure or other midbrain functions can reduce the ability to "pay attention". Other high sensor need for iteration will suppress the inferencing. If sensor data is "dull," inferencing will be prevalent.

Data from sensors are consolidated at or near the sensors, removing context provided by the environment being sensed. Context here is background or other information that relates sensed things to other things in the environment. Internal or relational context is established within the cortex in later steps. These steps result in data planes containing condensed information about the environment, which is then stored in memory. Almost all of this happens at the synapse and the next downstream neurons.

Certain data represents specific objects. This symbolic data is processed separately from other non-symbolic data. Words and language are symbolic data to humans. Faces are symbolic data in many animals. Symbolic processing evolved from a need to recognize offspring and family members, as well as certain other plants and animals. Human faces are symbolic in human brains. Human ears are not, however, even though they are as unique as faces.

Verbal communication followed as our ancestors developed muscles and face shapes that allowed for complicated audible symbols to be uttered. When the utterances were transcribed onto stone, papyrus, and paper, language was born, creating further dependence on symbolic processing. In today's humans, language by far overshadows all other forms of symbolism.

Again, separate input and output areas in the cortex exist for symbolic and non-symbolic processing. Here, I treat the midbrain (everything that is not the cortex) as a separate part of the brain that provides specialized non-symbolic data to the cortex and body. The symbolic processing sections or lobes are adjunct to the non-symbolic sections or lobes in that they only receive data via pathways from the non-symbolic sections and send data only to the non-symbolic sections. Early processing in the non-symbolic sections separates symbolic data for the symbolic section. Symbolic outputs are integrated in the non-symbolic sections to avoid or resolve conflicts.

Except for the above, and as Eugene pointed out, the symbolic section is otherwise independent of other sections. Energy management and the midbrain also play significant roles in intelligent thinking. Both are described later. The most important question, "How can you separate symbolic data without access to the information?" is answered later.

Fig. 2

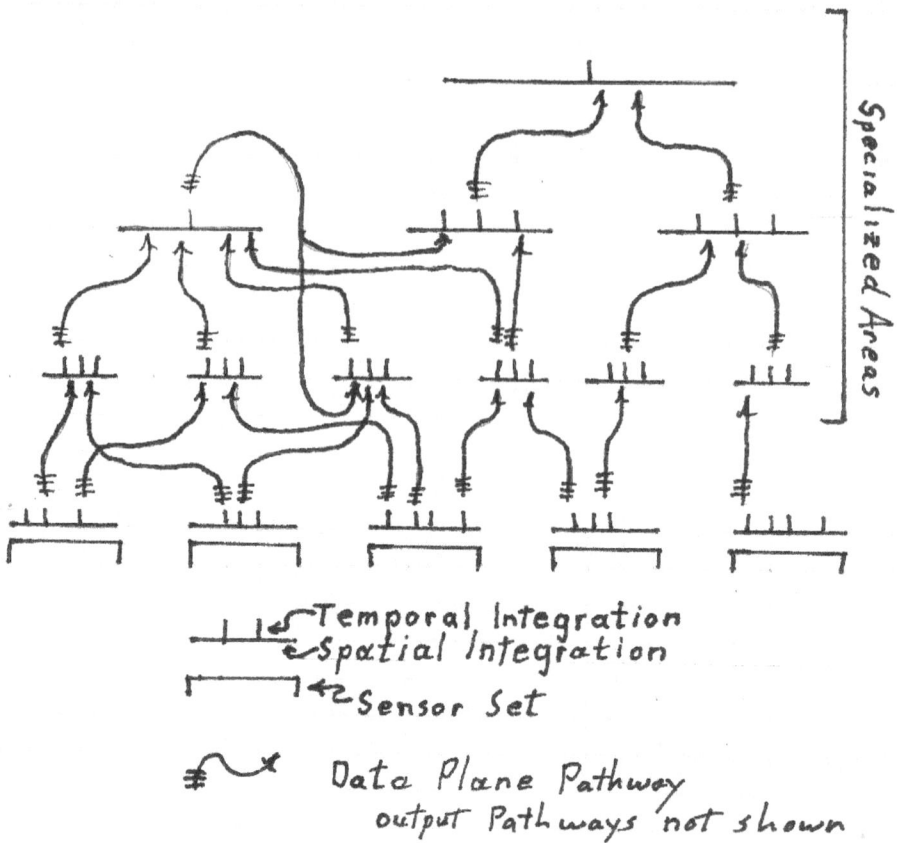

Temporal Integration
Spatial Integration
Sensor Set

Data Plane Pathway
output Pathways not shown

Cortical layers and data pathways

Figure 2 illustrates data paths within and among the symbolic and

non-symbolic sections. The symbolic area is represented on the left side of the drawing. Also shown are higher-level layers of specialized areas that have been created with data from input and other specialized areas. The higher-level layers cannot be created until the lower levels are working. The higher-level layers contain summarized information, with the top level containing non-symbolic behavior data that is general and persistent.

Sensor data is spatially and temporally abstracted, and the results are sent to the next layer for spatial and temporal abstraction and storage in a specialized area. At each layer above Layer 1, data from other sensors or specialized areas are included in spatial and temporal abstraction. Heard words and a seen speaker or words read aloud are examples of such integrated abstractions.

The layers above Layer 1 deal only with relational context data. If the data represents something specific that exists and persists in the environment, it is symbolic data and is processed separately from non-symbolic data. Symbolic data requires the very accurate matching of data planes to memory planes to avoid ambiguities. Non-symbolic data can be very rich in information content, and yet imperfect or partial matches with non-symbolic memory data are just as effective, sometimes even producing better results than exact matches.

Languages, music, and non-habitual motor skills are examples of environment-provided symbolic data whose processing is consciousness. Non-symbolic data processing is incorrectly labelled sub-consciousness.

The information in data planes traversing fiber bundles is stored in memory planes. Data planes leaving Layer 1 are condensed during abstraction. Layers 2 and above spatially and temporally abstract data on fibers from lower-layer outputs and store them in memory. At the top layer, all sensed data is combined in memory planes that represent general valid behaviors. The top layer is non-symbolic. In a lower layer, symbolic data is integrated (abstracted) into non-symbolic data. (e.g. The meaning of a sentence or a general perception of an entire book becomes non-symbolic.)

Iteration and Inferencing

The surface of the cortex contains specialized areas, such as areas related to speech. I will not attempt to reconcile the data that must exist in these areas with neurological labels presented in various textbooks. I will explain why later but, for now, I will refer to these areas only as specialized areas.

Bundles of nerve fiber connect afferent, or inbound, nerves to specialized areas, and each specialized area has its own efferent, or outbound, connections to other specialized areas, the midbrain or the rest of the body. All of these must have a purpose. For a given input, a specialized area memory will impose an output on its outbound fibers.

During general operation, two things are continuously happening in the cortex: sensory data is consolidated, stored, and transmitted to specialized areas, where they are integrated with other afferent fiber data; and outputs from specialized areas are conveyed to other specialized areas or to the midbrain and body. During both operations, data in transit is compared to memory, where it either matches something or it does not. Unmatched data is stored in memory, and arbitrary output is passed on. Input data planes are summed and rescaled, stored in memory, and passed on through efferent fibers. This step creates unique representations of data across time that then create new specialized areas and new outputs. This temporal abstraction is a critical part of the intelligent mechanism. This passing along of valid data among specialized areas is inferencing. Inferencing among the specialized areas in the symbolic section must comprise exact or near-exact matches and is called a procedure. Any ambiguity will affect the downstream inference steps. Jokes are symbolically ambiguous which causes the inference path to go astray. The ambiguous match is within a higher-level specialized speech area, and the output produced does not match the next input. After the mismatch, further inferencing leads to out-of-environmental-context results.

Non-symbolic inferencing is the predominate intelligent mechanism in humans. The symbolic areas send data to and receive data from the non-symbolic areas through a separate synchronization mechanism, so the non-symbolic areas receive the hierarchy of symbolic data. Recall and inferencing in the non-symbolic areas are not constrained to exact matches. Inexact matches provide the associative inferencing that Eugene incorrectly described. He asserted that random data planes were created, somehow, and compared to memory. The exalted data stage then generated whatever output was associated with a near-match. If the environment validated that output, the original inexact match would be stored in memory as a valid data plane or, more correctly, that data plane would not fade over time. His suggestion was to infer through layers of memory data planes and not between specialized areas. I can conjure up a plausible mechanism to generate the random data planes, but there is still a need for a mechanism to move the consequence or output to the antecedent parts of the memory plane. Since there are no data path switches in the cortex, this is problematic. I demonstrate a clever solution to this in the later sections of this book.

Non-symbolic inferencing is not part of consciousness. It continuously compares sensory, midbrain, and symbolic input with the built-up hierarchy of specialized area memory. Because the specialized areas are hardwired for output to input, inferencing can occur continuously. And because of inexact matches, allowed by a relaxation of match criteria, new associations are made with known valid outputs. This is subconscious, or nonverbal, thinking. "When I suddenly realized…", "I got it…", "of course…" are utterances frequently made when valid non-symbolic inferences are made and satisfy consequences passed on to and inferred within the non-symbolic cortex from the symbolic part of the brain.

One can follow a prescribed procedure within the symbolic cortex. Multiplying two large numbers is an example. The non-symbolic cortex cannot do this. It can only infer at the various levels of consolidated data and send the output data to the symbolic cortex, which learned to associate the data with its own symbols, or to the body, which responds accordingly. With sufficient data, the non-symbolic cortex is vastly more intelligent than the symbolic cortex. From the environment's point of view, the symbolic cortex is more intelligent in that it can interact with all sources of environmental symbolic information and enrich the environment's store of information by consolidating information, correlating it, and passing it back to the environment. The environment, from a single cortex's perspective, comprises all things it can sense, including information provided by other cortexes. Sorting out true and false information occurs over time. If the information persists in the environment, it has shown itself to be useful and is considered true. If it is not true or useful, it will eventually fade. This sorting is the only source of truth for any human cortex. There is no absolute truth standard, either in the cortex or in the environment (The goal in science and mathematics is to establish absolute truths.), nor is there any preloaded data or other information in the human brain, except for a few minor survival tricks known to the midbrain. The cortex, symbolic and non-symbolic, can hold any information as true or valid unless, through trials with the environment, it is shown to be false. Any cortex can build up stores of self-truths or beliefs through inferences that cause that person to act in a particular way, good or bad. The environment, through conversations about better beliefs, rule enforcement and inefficient or harmful results and actions, provides ongoing feedback about the validity of such beliefs. This feedback causes some beliefs to change and some to be reinforced. True human beliefs are developed over many years and stored as an amalgamation of context-dependent data in the highest level of the non-symbolic memory hierarchy. Consequently, they are difficult and slow to change. To further complicate things, the symbolic cortex data may not be in sync with the non-symbolic cortex data holding these beliefs. People can ramble on about things they do not really believe, or they can ramble on about things that they know they do not believe.

Higher-level data planes are information-rich and quite large. Exact matches with an entire data plane would be unlikely. Eugene's idea that the memory in the entire lobe is scanned synchronously provided a mechanism that would allow a small input data plane to match and fire a much larger data plane. Matching within a certain region of the larger data plane causes the entire lobe to fire, making the information in that memory plane available to downstream stages. This is regional matching.

3 DATA PATHWAYS

A sensor set or sensor system contains many individual sensor elements. The retina in the human eye contains rod cells and cone cells, which are individual light sensors or sensor elements. The output of each sensor element is moved to the next step in processing via a neuron and its axon. Each axon carries the analog output of a sensor element. Consequently, in this case, to convey the entire sensed image, a large number of neurons and axons are required. The values of the signals traveling along the axons at a specific instant make up a data plane. Data planes travel along groups or bundles of axons to the next set of neurons for the next stage in processing. Each stage involves spatial and temporal abstraction, the results of which are conveyed on another set of fibers to the next stage for integration with other sensor data or other specialized area output, until there are no more non-integrated specialized areas or sensor sets.

Figure 3

Temporal Integration Output to Next Layer

Memory Data

Exalted Data Stage

Input Data Planes

| Iteration Input From Output Section (Integration input) | Sensor Input or Lower Layer Temporal Integration Input | Inferential Antecedent Input from Specialized area Output | Context Free (**Direct to/from Memory) Consequence Output or Antecedent Input |

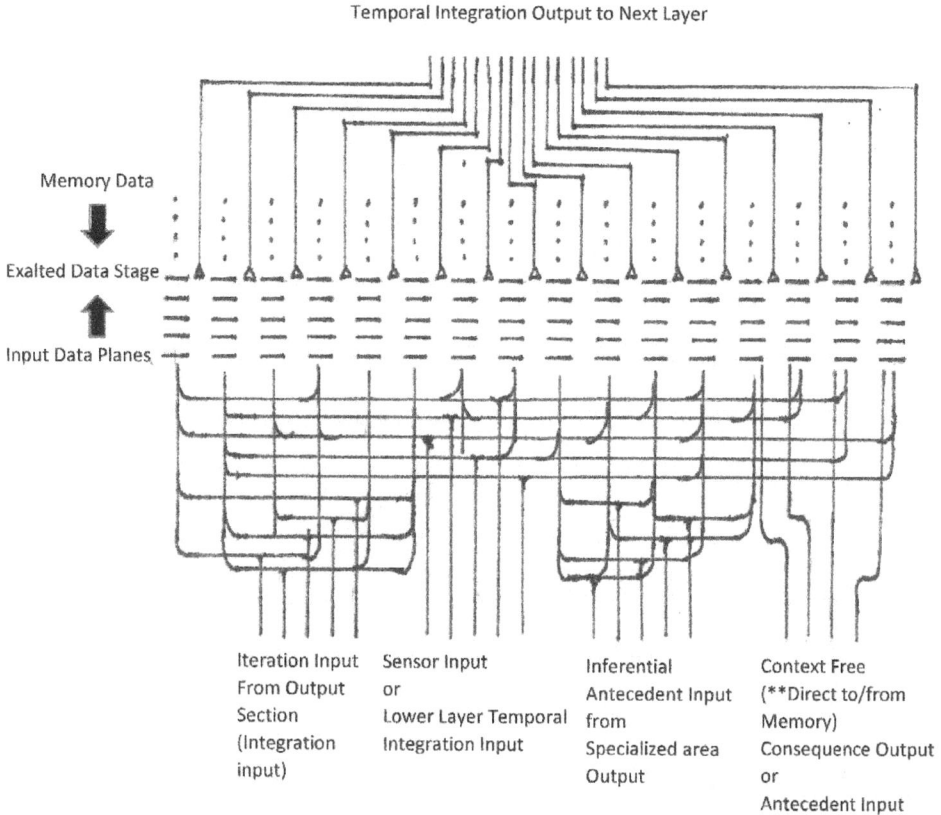

Figure 3 is an illustration of the fiber bundle connections associated with a specialized area. Sensor input, or lower-layer temporal abstraction input, comprises the fiber bundles carrying data to be integrated in this specialized area. It could include several sensor sets, such as vision and hearing, or it could include the outputs from several other specialized areas. There is a path between input and output sections that conveys the body's response to an output stream back to the input. This feedback loop should match the feedback

previously stored with the sensor data. Internal "talking and listening"

is an example. Inferencing occurs across specialized areas that also

have output fiber bundles integrated with various other specialized

areas. These bundles are integrated into the specialized areas to

maintain internal context. Eugene proposed a source of random data

planes to allow for associative inferencing, although he did not

postulate inferencing between specialized areas, which I do, but

described inferencing among memory planes. Without specific

physiology to create random data planes and a mechanism to

separate consequences and antecedents in memory, I cannot see

this model working. Instead, my model includes inferencing among

specialized areas, and the source of random data planes can simply

be fibers carrying data directly to and from memory (Since the fiber

connections were not determined by data values, the data planes are

context-free and can, therefore, serve as pseudorandom antecedents

or consequences, as shown on the lower-right side of Figure 3) or

more likely, inexact matches to current memory planes. At the top of

this illustration, a set of output fibers is shown carrying spatially and

temporally abstracted data to the specialized area in the next layer.

The layers are not physical layers of cortical cells but are logical layers defined by connections. Specialized areas and connections among them are not coded in DNA, with some exceptions such as input and output to the cortex. Early intracortical fiber connections are simply everything to everything, which is far easier to code in DNA. Most of these connections are culled in-utero and during early life based on input and output sensor data. While connection generation and culling can go wrong in many ways, it is generally argued that intelligence develops in early life through experience and learning which in turn culls unneeded connections.

4 SCALING AND CALIBRATION

One of the basic tenets of Eugene's construct is that the cortex is an analog system and, therefore, all components of the cortex must be precisely calibrated and have the same data value range. He left it unsaid that data elements in any data plane must have values that span nearly the entire range. Each data element value is the net sensor output energy for that sensor's duration and is not the peak or average value over that interval.

The scaling rule applies to all cortical transactions. It takes energy to maintain the scale range as data planes are processed. In certain cases, the underlying physiology may be unable to add enough energy to keep transactions valid. For example, as a data plane crosses synapses, the quantities of neural transmitters moving across the junction must be sufficient to maintain the scale range or scaling errors will occur, making downstream memory matches ambiguous. The general degradation of neurons' metabolic systems, or even larger systemic issues such as low serum PO_2 levels, could cause such problems. I suspect that certain psychotropic drugs also cause such a shift in the scale range of the data, causing memory mismatches that are planned or unplanned departures from normal cognition.

5 SPATIAL ABSTRACTION

Eugene did not explain the lateral reorganization of the data network completely, although he hinted at the missing pieces. He claimed that data streams from individual sensor elements or other data pathways must be related to each other and, in the cortex, a physical construct is needed to do this. This is referred to as the *LATERAL TO AXIS REORGANIZATION OF THE DATA BUSES THAT THEY ARE CARRYING* in his book and as "spatial abstraction" here. Eugene simply claimed that about 50% of inbound axons must branch, using dendritic growth, to other outbound neurons and axons to ensure the data plane elements are sufficiently integrated. He does not account for the appearance of specialized areas on the surface of the cortex.

An input data stream from an individual sensor element is not valid unless it is associated with something, generally the rest of the input streams from the other sensor elements. The dendrites of neurons conveying input data provide such context by connecting to many other cells that form the next downstream stage. Each of the inputs from single-sensor elements (actually multiplexed inputs) contributes some part of its analog sensor data to perhaps 50% of the cells in the next stage. Eugene suggested that these connections were formed by energetic dendrites seeking a ground or cell upon which their energy could be dissipated, i.e. where their neurotransmitters could be absorbed. The dendrites grow towards neurons capable of dissipating the added energy. Eventually, the nearby neurons are saturated, and the dendrites grow towards neurons further away. This is probably how connections are originally established in-utero or shortly after birth. This structure leaves the idea that areas on the cortex related to seemingly recognizable information, unaccounted for.

Broca's area, located in the left hemisphere of the frontal lobe, is an example of such an area. Its size is related to early language development. If an adult learns a second language proficiently, his or her Broca's area does not grow larger; rather, a new area forms next to the original area. Cells in the newly added area are not new neurons but were previously used for other purposes. This can only happen if the lateral reorganization is data-driven. Many of the downstream neurons, especially those that are at the periphery of a different specialized area, are capable of more connections and so are underutilized. It is these underutilized neurons that can be repurposed as data sinks for new but common data, thus accounting for why learning a new language increases Broca's area asymmetrically.

(There are many named areas on the surface of the cortex. Eugene used several but, to avoid the pitfalls associated with relating behavior to cortical areas, with just a few exceptions, I elected to refer to all such areas as specialized areas.)

It is possible that, initially, the spatial abstraction mapping begins as either chemically driven connections or energy-driven connections that use background data or noise. This likely begins in the womb.

Spatial abstraction mapping development begins when data sensing begins and continues over a person's first 10 or 12 years of life, generally ending around the time the skull stops growing. However, the development of new mappings can continue into later life. It appears to be more difficult to change the mappings as the brain ages, but rehabilitation from brain injury demonstrates that new mappings can be created. If a mapping is damaged significantly, the data passing through that abstraction stage will not match memory, and the memory data will be lost forever.

It is important to mention here that common scaling and spatial abstraction constructs also solve the problem of integrating data from different sensor types. Visual data planes and related audible data planes can be integrated in spatially abstraction mappings so that all of the information is spatially related and represented in a larger data plane in the next stage or layer.

6 TEMPORAL ABSTRACTION

Eugene also suggested that memory scanning synchronization with sensor input streams is accomplished with a somewhat standard proprioception identifier at the beginning of each related sequence of data planes, in both memory and data planes from sensors. Problems with such synchronization are compounded by a problem that he recognized in earlier editions of his book, but which was not included in the current one. That is, when first perceived, a memory stream enters memory with the oldest data plane stored first and apparently retrieved last. Memory appears to be read backward. (As a solution, he also postulated a linkage between data planes to maintain their order during sleep and other memory clean-up procedures. Data P_2 has a pointer to P_1 and P_3, so its position in data is independent of its position in memory. This became untenable and, later, unnecessary.) Eugene eventually solved the first-in last-out memory problem by providing two copies of memory: one being sensory input and the other being sensor data memory. Regardless of how they are stored, they are entered together.

Other problems occurred in his sequence of data streams. The oldest plane in the data stream must be related to newer planes, otherwise, data will not have context within the stream. Furthermore, as output energy expenditure units reported to the cortex by the midbrain are included in new memory entries, there must be a mechanism to sum these energies for allowing later comparisons in abstract data planes. Eugene idea provides no way to summarize energy expenditures associated with a scenario. Matching the memory stream with the input stream does not solve these problems. Therefore, his theory of simply reading out the tape does not work.

Another aspect of his construct is the idea that a data planes transmitted to the next region or lobe are the difference between input data and memory data, as compared in the exalted data stage. He referred to these differences data planes as a "language specific to the lobe." These data planes do not comply with the scale range distribution constraint. The differences in a close match would be nearly zero across the output plane, making a valid match produce unusable data for the next abstraction or output instruction. It would be unusable because it does not meet the scaling range requirement. Rescaling is required. He did not account for this problem.

Trying to resolve his inferential mechanism, the need for temporal abstraction, such as data plane sequencing, and the limits imposed by the physiological mechanism were frustrating, as were separating non-symbolic inferencing and associative thinking without continually defaulting to human mental behaviors that have little or nothing to do with this subject.

Recently, I developed a simple and elegant solution for all of the above issues. Such a solution must ensure that data is scaled within the entire scale range.

Temporal abstraction occurs in the same mechanism as Spatial abstraction. As values from a sensor or other source traverse the spatial abstraction fibers, they lose energy or data values based on the geographical distance the signal must travel. Those fibers close to the downstream neuron lose less than those that extend to the furthest reaches. If the arriving data planes do not create a center loci of values high enough to fire the stage, simply waiting and summing more data passing through the fibers will quickly cause the scale range to be exceeded in a small area near the center of the set of fibers and cause the entire stage to fire, storing the summed data. If the system is overdriven overtime, data values accumulate too fast, the spatial integration layer will adjust the pathways (disconnect some higher summed values pathways and make new connections to lower summed values pathways) to better maintain the overall scale range. So, many data planes will have been summed into one and the scale range will have reset the data value losses due to the longer paths of some fibers. Additionally, the higher values at near the center of the set of pathways will create a specialize area by causing more connections to be created around the center.

The loss of signal strength over the data paths is key to making this mechanism work. There are two variations that ensure the losses are sufficient to create an epoch defined by the data. The common mechanism is a gain control or feedback mechanism as postulated by Eugene for sensors. If the signal strength is increasing rapidly, the gain control increases resistance in the data path to throttle the increase. The gain amount is stored alongside the data. The throttling means that more data planes are accumulated before the firing threshold is met. Since the gain values are stored, the system can learn to refine the gain as the same or similar sensor data is encountered. Refining here means adjusting the epoch to ensure it is consistent and contains a complete sequence of related data planes.

The second variation is less common and is needed for longer epochs, particularly symbol extraction. Eugene said the data value differences computed in the exalted data stage and sent along output fibers are a language specific to the lobe. The differences values for sensor data matching memory data would be very low and would never meet a threshold that would fire the next stage. The values must be accumulated or boosted to meet the scale range requirement. Those values can be accumulated much like the first variation but without a gain control mechanism. Because the data values are so low, a substantial amount of time would be required to accumulate enough data planes to reach a firing threshold somewhere in the exalted data stage. This mechanism works because data plane sets of identical sensed information are not identical, there are slight variations and noise in all such signals or sensed data. The spatial integration layers ensure that inexact matching of this data for a given epoch is close enough to make it unambiguous. I suspect this only works on long fiber tracts.

Several amazing things have occurred. A series of data planes that are temporally related are now integrated into one new data plane. The new data plane is larger than the input planes and contains many times the amount of data. This plane contains information provided by the series of planes used to create it; it is unique to that series of data planes. The data is abstracted across environmental time. This only works because the inbound information is context free, i.e. it has passed through spatial integration. Scale ranges are adjusted and maintained (TA is an energy pump). Specialized areas and Temporal Abstraction are accounted for.

Over time, these new spatially and temporally abstracted data planes, if they are valid, build up in memory and form the input to other specialized areas for further integration, abstraction or output. Sensor data will never match a temporally integrated plane, so the sensory memory planes used for the integration must remain in memory to match future sensory data.

This is a very elegant solution in that it is easily accommodated within the physiological apparatus available. The only new requirement is that the appropriate neurons can delay firing until the energy levels of some of the comparator results meet the scaling requirement, but even this is not new.

This also makes the temporal abstractor the energy pump for the data. Data values are not voltages or currents but are the integral of the voltage signals for a given epoch. The physiological mechanism may simply compound neurotransmitters' quantities based on the voltages during the epoch. (Such regulations, based only on an instantaneous membrane voltage, would reduce the system's resolution to the point that it would not work.)

The other truly amazing benefit to temporal abstraction is that it solves the synchronization problem by nearly eliminating it. There is no longer a need to find the starting point of a related series of data planes in memory because the information they contained has been recoded into a single data plane. Once past the initial sensory input comparison to sensory memory, synchronization is effectively accomplished by matching a memory plane to an input data plane. Any set of planes that needs to be related, regardless of its source, is related in a next-stage temporal abstractor. There is a difference between the sizes of the planes that are abstracted and the size of the abstracted plane, with the abstracted plane being larger (i.e. it has a greater number of neurons to conduct the data). This data spreading is done by the same mechanism that performs the spatial abstraction (i.e. lateral reorganization).

Temporally abstracted data planes are spatially integrated with other temporally abstracted data planes from other areas to create new data planes that include information from all of those areas. It does not take many integration layers to accumulate all sensory data and specialized area data, including the abstraction of symbolic data, into what is most likely a set of top-level data planes that are very broad but not particularly deep, all of which are related.

This ability to relate and abstract all sensed data is the mechanism that allows partial and inexact matching as described later.

6A TEMPORAL ABSTRACTION OF SYMBOLS

Without access to the subject matter, how can the cortex separate symbolic and non-symbolic data? (Earlier I warned of the pitfalls associated with allowing human behavior to encroach upon the logic of cortical data processing.) This problem stymied me for a long time because of the behavior issue. Thinking about a symbol as being a symbol is a behavior. In the cortex, a symbol is only a symbol because I describe it that way. Symbols are still a cortical data construct but they are actually defined by the environment.

In the Exalted Data Stage, incoming data is compared to memory data and the streams of differences continue if good matches are maintained. The differences between input and memory streaming data values are computed in the EDS and conveyed as analog signals to the next spatial abstractor where they are summed until the Scale Range is met for a few elements. If the EDS differences are smaller over time, the epoch is much larger than normal. So large in fact that it includes all of the time necessary for the sensor systems to scan the entire item in the environment (e.g. a face, a group of printed words, a series of sounds, etc.). The data plane constructed during that epoch is the symbol. The fiber tracts containing the data value differences convey the data to the symbolic sections of the cortex. Long fiber tracts require long epochs to prevent early firings of the ESD and thus ensuring the entire symbol is abstracted. Thus, symbolic sections of the cortex become physically separated from the non-symbolic sections. In the course of fiber tract development, long epoch signals may also traverse shorter fiber tracts and be stored as incongruous data, later faded from memory. (These fiber tracts and the temporal abstraction of symbols are likely the most important evolutionary change in the human cortex.)

7 MEMORY STORAGE AND RECALL

Sensor data is stored in memory after the first SA/TA conversion. Streams of input data that do not match memory needs to be stored in memory for next stage temporal abstraction but that abstraction requires a segment in memory that matches, or nearly matches, the input stream. Eugene included a second copy of sensor memory in his model to fix the ordering problem. (i.e. the last stored memory stream sequence would be inverted when compared to the next encounter of that same data so the inverted stream is sent to different memory; first out is first in, in that memory thus correcting the order problem). It looks to me like the second memory is a buffer that reorders chunks of the sensor data stream and feeds it back into sensor memory if there is no match in memory to the current sensor input stream. It buffers a copy of the sensor data stream during memory search and if the search fails, provide the corrected input stream into memory. This complexity solves the critical problem of reducing sensor memory capacity demands because data already in sensor memory is not re-stored, reordering the data streams and storing a copy of unique sensor data on the first exposure to it. The physiology is not unique to sensor memory in that it allows input or output memory in any layer to store streams of data continuously. There maybe a cutoff ESD firing to signal the end of an input data stream probably caused by other input data, such as a halt in eye scanning.

Now, with everything in memory, temporal abstraction can occur accurately. Sensor or any input stream can synch up with memory scans and the differences in every value between the two data planes is conveyed via a fiber tract to the next SA/TA stage where they are summed for the duration of the input stream and stored. This action does fire the ESD, entering a single data plane into memory, searching memory for that same data and conveying the differences to the next SA/TA layer.

Input memory and output memory are independent. They are scanned separately. Eugene proposed that they just resynched or were always in synch so that when an input match was made an associated output could be made. Computationally, this seems unlikely. The output memory is spatially and temporally abstracted and that data is sent to the next integration layer where a match will not restart the output memory scan but instead let the output continue to stream. If the integration match is poor, resynching is done by starting the output memory scan over. With this idea, it is now possible for the integration layer to find an integration data plane that matches its input side and then allow output memory to scan until the memory values present by the integration layer matches something. The easiest solution for streaming output data is to have a single integration memory plane associated with the entire set of data planes in the stream. Output streams as long as the associated integration memory data plane matches the output data planes.

The compression effect of Temporal Abstraction reduced energy and storage demands in the cortex and creates layers of abstracted data.

These abstracted data planes permit exact, partial and inexact matches to input data. The partial and inexact matches will also minimize output energy requirements because the closest match will produce an output that is likely the most efficient. Specialized-area-to-midbrain fibers send quantitative information about the match quality (the stage firing) to the midbrain, which responds with a positive reaction (sometimes called reinforcement), the effects of which are then stored in memory as part of the new copy of the successfully matched data plane. The midbrain reaction itself cannot be stored in memory. (Pain, pleasure, fear, etc. are midbrain reactions that cannot be stored or recalled.) The overall result is the reinforcement of efficient and accurate scanning, making a next match more likely. The positive reinforcement is pleasurable on a very small scale; a little "squirt" of happiness anytime there is a strong cortical match. This reinforcement accounts for a great deal of human behavior. Finding familiar things is important for survival. Not rewarding poor matches prevents aimless inferencing or iteration, that is, catatonic like behavior. A simple demonstration of this is to look through a collection of old photographs of people familiar and unfamiliar to you. The match sensations are more than just recognition.

Finally, inexact matches, i.e. any input that almost matches memory, will produce something between near-zero differences (a full match) and near-maximum differences (a poor match). Many such inexact matches will produce ambiguous results that are unusable. Closer inexact matches produce an output that may work. They may produce workable inferences, or they may produce an output that may be validated by the environment. The ability to make inexact matches this way solves the most onerous artificial intelligence problem: sensing and operating in an environment without prior knowledge of that environment, something algorithms and heuristics cannot do.

8 INTRODUCTIONS TO THE INTELLIGENT MECHANISMS

Iteration is a conversation, symbolic or non-symbolic, with the environment. A certain input generates an output to the environment directly from memory through one or more layers. Ideas, sentences, and words are recalled and uttered as habit, as are walking and running and most other motor activities. The cortex senses its own output. This self-sensing or listening forms an internal iterative loop, which contains little new information. The output to the environment may elicit new or already known information from the environment. The new input is stored and matched to whatever similar experience the receiving cortex has in memory or is ignored or stored and used inferentially. The known information results in memory matches and more output to the environment. This iterative intelligence is far older than humans. The symbolic part is the exception and appeared as languages and inscriptions became prevalent.

Inferencing is another internal conversation. Fibers between specialized areas in the cortex convey information to other specialized areas, where memory matches, exact or inexact, are made and stored in memory. Output from that specialized area is then conveyed to another such area and so on. When exact matches are necessary, these internal loops are procedural mechanisms. When inexact match opportunities present themselves, these internal loops produce associative inferences. Inexact matches are always possible simply by loosening the match criteria. Loosening the match criteria make data planes appear to contain the random data that Eugene alluded to but did not explain. Associative thinking creates new information and is regarded as the intelligent mechanism bestowed only upon humans. The rise of associative thinking required physiological changes that appeared about 20,000 years ago and became historically prominent about 10,000 years ago.

9 ITERATION

Eugene's definitions of associative thinking and iterative thinking are a little different than how mechanisms must work. He acknowledged that data can flow in circular paths between lobes and within lobes or through interaction with the environment. I will call this circular flow "inferencing", and I will illustrate it later.

Iterations occur between memory recall and the environment and are generally trial-and-error outputs until a set of data planes produce outputs that are reinforced by the environment.

Matching between specialized areas up through sensor data areas and down through output areas is known as iteration. Matching, inexact or otherwise, among specialized areas at the same or other levels is inferencing and requires different connections, as described above. These direct-to-memory connections explain the purpose of many intracortical fiber connections, provide context-free data that satisfies the randomness requirement for inexact matches, and do not interfere with context dependent data planes since they are just an ancillary but critical non-abstracted part of these data planes.

Labels regarding brain function and thinking are environmental artifacts and do not specifically define brain functions. "I remember," "I realized," and "I thought about it" are examples of such artifacts. Each can refer to successful (produced an output) recalls, associative recalls, or inference results. Most of these labels refer to observed or learned behavior and not to specific cortical functions.

10 INFERENCE

Eugene did not clarify his idea of data flow in the cortex. Output from a lobe may be conveyed to the next lobe where, as an input, a match is attempted, but it is not clear where that output goes and whether or not it is direct output or requires further processing.

There could be many steps in a symbolic inferential loop. If everything matches well, it becomes an internal "conversation", with few or no new trials created. This symbolic loop works only because the matches are highly accurate. It serves to reinforce what is already in memory and, possibly, creates more combinations of symbolic sequences. This symbolic inferencing is the offline process Eugene described as using the insular lobe to resync with the real-time, non-symbolic, cortex. This resyncing mechanism is a buffer between the two systems and allows the symbolic inference system to input to and accept output from the non-symbolic system.

Eugene does not postulate a non-symbolic inferential loop, but I do. Much of the data passed around in the cortex is not related to symbols used in the environment but is comprised of values that have meaning within the contexts of a specific cortex. Generally, these would be integrations of sensory inputs: sound and sight, taste and smell, midbrain output, and proprioception and integrated symbols. Spatial abstraction occurs in layers, starting with specific sensory data being extracted, stored, and passed on to the next step. It is then temporally abstracted and passed on to the next layer for integration with other data. This continues until all sensory and specialized area data have been integrated. The result is a set of layers of data planes that embodies all behaviors, habits, and abstract representations of all sensed things. The highest-layer data planes are very broad and fewer than lower-layer data planes.

Inferencing within these layers can produce new results because associative matches will very likely generate something that will match in downstream or lower-level layers. Any new results must be validated by the environment. This can be done indirectly through further inferencing with previously validated data. This is the newest intelligent mechanism in the cortex.

The mechanism for transferring data between the symbolic and non-symbolic lobes (the insular lobe) must be robust. Heard or read ideas must be integrated into the non-symbolic data planes so that all beliefs can eventually be abstracted to the top-layer planes.

As an example of this transfer, a student may study a homework assignment so the words or symbols can be recited in class or in a test. Further study, which would be further symbolic abstractions passed to the non-symbolic memory through inference loops, could result in the incorporation of the new material into related non-symbolic data planes. ("Oh, I get it!" is an utterance that is sometimes an indicator of successfully relating new information with stored information through inferencing) The student that doesn't "get it" may pass a test but will miss the questions in which the idea must be reapplied in a different environmental context.

Classical inferencing between data planes uses defined antecedents and consequences as indices and links to the next data plane. Consider these three data planes (shown vertically):

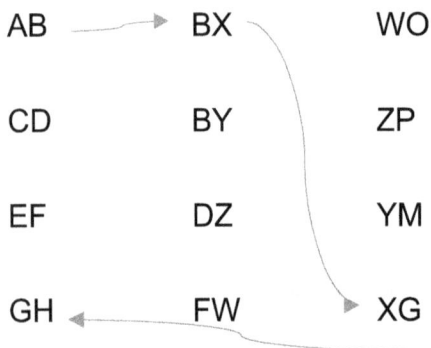

AB →	BX	WO
CD	BY	ZP
EF	DZ	YM
GH ←	FW	→ XG

The first letter in each pair is an antecedent, the second is a consequence. If A is sensed and then matched in the first data plane,

B will be transmitted and compared in the second data plane, where the first encounter with B in memory will produce an X as a consequence. The X will produce a G as a consequence in the next data plane, which can then be fed back to the first data plane to produce an H and so on.

In the cortex, matches can be exact or inexact. In the case of associative inferencing, there is sufficient apparent randomness in the context free data or almost any memory data for that matter, that the antecedent forces inexact matches which creates new information. In the case of symbolic inferencing, matches must be precise. (Exact matches require more metabolic energy, so tired brains may not produce valid symbolic results, e.g. "I can't think straight".) A symbolic inferencing path is consistent and is a reusable procedure. A non-symbolic inferencing path is inconsistent and utilizes inexact matching. It adds new and somewhat unique data planes that may be valid to memory.

Inferencing occurs between specialized areas and is considered thinking. Iteration is output back to input and is also considered thinking. Symbolic inferencing and iteration are considered consciousness. Non-symbolic iteration is real-time behavior or scenario recalls (things like navigation and daydreaming). Non-symbolic inferencing is figuring things out through associations. Figuring things out non-symbolically is not part of the consciousness belief habit and is sometimes called subconscious; super-conscious may actually be a better term.

Taken a step further, symbolic inferencing is also the execution of steps or a procedure. You learn X, you learn Y, and you learn Z, and you learn that X, Y, and Z are each associated with A, B, and C. You also learn that X and Y form XY, and Y and Z form YZ. Assume this is symbolic data upon which the procedure will act.

Then, you learn a procedure in which A precedes B and B precedes C. Temporal abstraction abstracts the procedures to a higher-level data plane that contains information that A precedes B which precedes C. This abstraction is just a "handle" which matches the input that initiates the procedure.

The output consequences from the higher-level plane are A and a step to treat input X as an antecedent through different fibers to a different specialized area, where it would match an X in memory and produce an output of Y. They would look something like this in memory:

(Consequences are underlined) (Input and output are asynchronous)

(Data plane instances in Layer 2, Layer 1, Area 1, and Area 2 are in the same data stage)

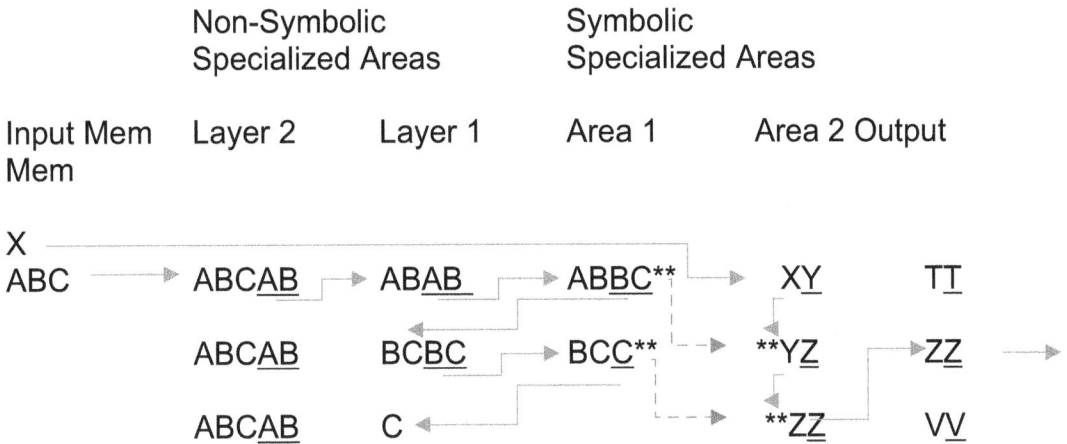

	Non-Symbolic Specialized Areas		Symbolic Specialized Areas	

Input Mem Mem	Layer 2	Layer 1	Area 1	Area 2 Output
X ABC	ABCA<u>B</u> →	ABA<u>B</u> →	AB<u>B</u>C** →	X<u>Y</u> T<u>T</u>
	ABCA<u>B</u>	BC<u>B</u>C →	BC<u>C</u>** ⇢	**Y<u>Z</u> → Z<u>Z</u> →
	ABCA<u>B</u>	C ←		**Z<u>Z</u> V<u>V</u>

This works because a match of consequence statements and an

inexact match of the context-free data will cause the entire data stage

to fire. (Data is conveyed on fiber bundles (solid arrows). Context-free

data (dashed arrows) contains no information beyond being unique

identifiers for memory retrieval in a different specialized area. The

input or antecedent data and the context-free data must match to fire

the data stage.) There are no switches in the cortex, data planes

follow specific paths. Data must be in memory. If it is not there,

appropriate earlier learning did not occur. To be consistent, the data

inferencing in Area 2 is actually iteration using an internal feedback

loop.

This procedure of learning and execution allows for language beyond

grunts and shouts. It also allows for learning other procedures such as singing, oration, and physical tasks with many steps and so on. If a procedure is executed frequently, the output is stored as a stream in non-symbolic memory, and the procedure is not used for that output thereafter. This is incorrectly called the "muscle memory" for physical tasks. For example, in golf, a swing requires one to have knees bent, be square on the ball, have one's forward arm straight, etc., which eventually becomes a swing with no thought about the mechanics needed. Practicing a song or speech until it is memorized is another example. Other procedures are used infrequently. Solving a long division math problem is an example. 8/4 is usually directly recalled from memory. 1365/15 requires many steps and intermediate symbolic results.

A self-realized procedure is a plan. (A learned plan is still a procedure. All plans and procedures can be executed in the future). Being able to figure out a plan to accomplish a goal gave early humans a tremendous advantage, whether it be to find food, defend one's family from enemies, or develop a language that allowed for learning things that were beyond personal observation.

Imagine, in the same illustration, that the matches in Area 1 or Area 2 are inexact due to the randomness of the context-free data on the fiber tracts, as illustrated with red arrows. During execution, the procedure would not follow the learned or prescribed path. It would wander as new associations were made and stored as output. This is an example of associative inferencing, the results of which are not learned and retained information and, therefore, may not be valid.

Non-symbolic inferencing can retrieve lost symbolic memories by associations with related memories. Relating a name to a heard song or a name to a familiar face are examples. Matching the latter consciously through symbolic recall or symbolic procedures sometimes fails, i.e. you know that you cannot recall the name. Shifting the conscious thought to something else can leave the non-symbolic section inference mechanism searching for something related to either part, the name or the face. Unexpectedly, you remember the name because a subconscious scenario or association included the name and face as part of something else. This is a powerful mechanism for restoring weak or non-retrievable memories.

Temporal abstraction predates symbolic inferencing (e.g. speech) by about 10,000 years. It appears that both played a role in the average cortex volume decrease that occurred about 10,000 years ago.

Inferencing is the intelligent mechanism in humans. Iterating is only as intelligent as the responding environmental players.

In summary, the inferencing mechanisms:

1) Refresh memory by finding lost or irretrievable memory planes through association with related memories when iterative recall does not work correctly. (Finding your car keys by retracing your steps mentally after failing to recall where you left them, for example. Another example is failing to recall the name of something and suddenly remembering it after you stopped trying to recall it.)

2) Reinforce memories by making redundant copies at levels higher than sensed data levels. Multiple copies improve the likelihood of a particular data plane being recalled. (Traumatic experiences result in many copies of similar data planes, for example). Using inferencing to make redundant copies in this manner ensures a greater likelihood of that information being recalled. This is a clever use of the inference mechanism by humans to maintain focus on important things. Overdone, this kind of

inferencing is called worrying and can create anxiety responses in the midbrain.

3) Integrate stored sensed input that is not related to higher-level memory planes. This input is out of context until it is stored with other related things at the top-level specialized area. (Context here is not spatial abstraction context; it is the relationship between stored information in the cortex.) It can still be used but is not related to fundamental or habitual beliefs. People use this out-of-context information frequently; usually successfully, sometimes not as successfully. (See the section on measuring intelligence.)

4) Execute learned step-by-step procedures, such as speech construction or deconstruction or long division, as described above.

5) Associate new data planes to similar stored data planes, creating new information that, depending on the fidelity of memory data, may be valid.

Points 4 and 5 are intelligent mechanisms that evolved in the human cortex, generally due to humans' creation of environmental pressures.

11 SYSTEM CONSTRUCTS AND CORTEX DEVELOPMENT

Figure 4 provides two example structures for cortical pathways and abstraction layers and a rough illustration of the ages at which the paths and levels develop. With exposure to an information-rich environment, i.e. a variety of useful uncommon information, specialized areas are formed, which then allow for integration with other specialized areas in the next layer. Like Eugene, I suspect that there are only a few layers, with the actual number being entirely dependent on the number of lower-level specialized areas: More sensory data produces more higher level specialized areas.

It is easy to see that a structure and pathway deficit in early life, possibly due to physical issues such as nutrition or, more likely, a lack of exposure to new information at the proper time and for the proper duration, will substantially limit higher levels of integration and, therefore, functional intelligence.

Learning to count toy blocks integrates auditory, visual, and motor skill specialized areas, which are precursors to specialized areas for other basic mathematical concepts. Without the former, the latter area is not formed, so certain mathematical concepts are not available. If such an area could be located on the surface of the cortex, it would be very difficult to assign meaning or purpose, or even an accurate name, to the area. Again, there are very likely many specialized areas related to mathematical concepts that are then used to form other specialized areas for higher-level concepts. These areas and their associated data are non-symbolic, even though mathematical procedures are symbolic (e.g. any equation).

School systems are gradually optimizing age-specific curricula to improve students' overall cortical structure within the limits of common classroom teaching and intelligence measurements. Unfortunately, these curricula (or budgets) are not flexible enough to detect and optimize different learning needs attributable to structure and pathway deficits formed prior to school-going age. Learning not only adds content to memory—it changes pathways and abstraction layers.

Figure 4 illustrates this problem. The top drawing illustrates a structure with limited integration due to limited uncommon data at the sensory level, which limits integration at the next level and so on. The lower drawing illustrates a more robust structure with an additional integration layer.

Early learning with false or contrived data such as movies and fictional literature will eventually lead to higher level data abstractions that are not valid or at least not completely valid. This, in turn, can lead to behaviors that are inconsistent or incompatible with reality which in turn causes internal conflicts in the cortex. Real world experience in early life is vital in avoiding this problem.

Figure 4

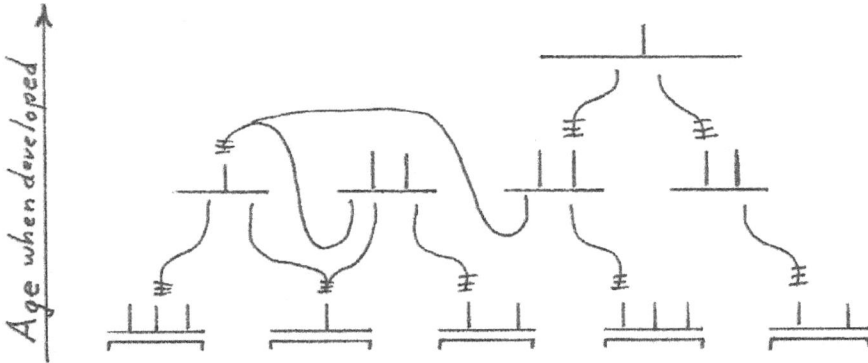

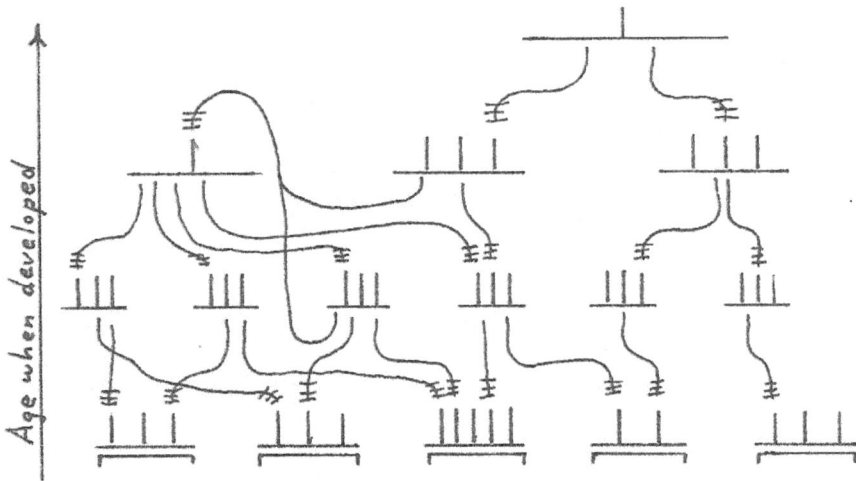

Temporal Integration and Memory
Spatial Integration
Sensor Set

Data Plane Pathway (Typical)

12 MIDBRAIN INPUT TO THE CORTEX

I broadly define the midbrain as everything that is not the cortex or sensors. This would include the lower-brain components, sometimes referred to as the "primitive brain". I only discuss it here because it plays a far greater role in intelligent behavior than Eugene theorized. However, my explanations are simplified.

Originally, the cortex was the memory service of the midbrain, giving the host organism an advantage by allowing it to learn what it could about the things that it needed to survive: terrain or ocean topography; the shape, size, and smell of predators; and the shape, size, and smell of family members. Without a store of experience, the midbrain could only react to immediate events with preprogrammed actions or instincts, as some creatures still do.

The midbrain samples all input data planes and reacts to these inputs with inborn (instinctive) responses. The cortical memory service allows the midbrain to keep and reuse the responses to sensory data that prove to be useful.

The midbrain is the source of a variety of functions necessary for sustaining life and procreating. Some, like breathing, occur continuously and are easily observed. Most depend on environmental stimuli or body sensors and result in chemical or electronic signals to the body or cortex. Love, anger, fear, flight from danger, hunger, and pain are examples that still exist and have these labels assigned to them. The labels are artifacts that are useful within the environment but do not adequately characterize the functions.

Pain and pleasure seem to be absolute sensations in the midbrain (i.e. they cannot be constructed from other functions or sensations). Very early in life, an infant quickly learns other habits associated with pain and pleasure, such as avoiding and mitigating pain and desiring, attracting, and maintaining pleasure. These cortical habits become the basis for many other habits that constitute human behavior. They are included in most, if not all, specialized areas and the data planes stored in these areas. In many cases, especially in early life, matching these areas in a data plane will fire the entire stage. Much of the behavior in humans can be traced back to habits created from early learning of the need for pleasure or the avoidance of pain.

The "pleasure" midbrain sensation stimulates the cortex to form the habit "want more pleasure." This habit, if effective, will cause another higher-level habit, "want more", to be created. Any particular class of "want more" things that is successfully acquired will create another habit just for that class of things. The idea here is that rudimentary habits allow for the formation of higher-level habits that then cause the formation of even higher-level habits. Undoing this to change destructive or antisocial habits is difficult.

The midbrain can control human behavior in certain instances. In most cases, the modern midbrain sends information to the cortex, and the cortex then provides the output. This includes rewards and avoidance sensations and all other midbrain-created sensations that are sensed in the midbrain. A cortical representation of a sensation's event or history is stored in memory, but the sensation itself is not. You cannot recall the sensations of pain or pleasure.

Behavioral descriptions are fraught with midbrain activities: midbrain functions initiated by sensed data, hormonal data or internal midbrain data, and functions initiated by the midbrains' reception of cortex outputs. In fact, the midbrain initiates most behavior and relies on the cortex for output that has been validated by prior experience. This includes hunger, fatigue (energy depletion and insufficient sleep), sex drive, illness mitigation, and pleasant and unpleasant encounters, to name a few.

The midbrain does seem to learn, which is the operant conditioning that Eugene contested. (I think he was assuming that the reported operant conditioning was conditioning or acting on the cortex, but I did not get the opportunity to discuss this with him.)

California condors are godawful ugly birds that only another condor could love. They do, and they mate for life. Imagine a few male condors on a tree branch looking at a small group of female condors on a branch across from them. Something happens in the midbrain of one of the boys and one of the girls, and they are bonded forever. I realize the mating ritual is more elaborate than that but, in the case of the condor and many other animals, a change in the midbrain connects the two partners for life. It is a chemical change, and maybe a neuronal connection change, that is permanent. Humans experience the rudiments of this midbrain change and call it "love". The follow-on bonding behavior between humans is learned cortical behavior, although separation anxiety and other emotional sensations associated with the bonding are examples of midbrain-instigated but cortically seated behavior. Thus, Pavlov's operant conditioning does occur in the human brain, but it is separate and different from behaviors that the cortex learns from the environment.

I treat the midbrain as a black box, with data paths to and from the cortex, to separate the midbrain portion of behavior from the cortical portion. There are ongoing iterative feedback loops between the cortex and midbrain, much like the feedback loops within the cortex and between the cortex and the environment. While it seems as though there are a large number of sensations, it only takes a small number to account for most midbrain-to-cortex data. Flavors, for example, are sensed in the midbrain as a combination of data from four or five receptor types on the tongue. The tremendous variety of food that we eat can be described with just four sensations. There are data paths to the cortex for flavors, smells, hunger, and other cortex-recognized midbrain sensations. There are some other midbrain sensations that are not generally recognized but are specific, critical inputs to the cortex. Energy expenditure is an example. The midbrain sends data, indirectly related to energy expenditure, to the cortex. Body temperature, blood pressure, heart rate, breathing rate, and others are sensed in the midbrain and sent to the cortex as sensed data. The integration and abstraction of this data is done in the cortex and has no labels or symbols, representing only the energy expended on the current activity. Later, I will propose that the temporal abstraction of this data provides humans with a unique ability to plan activities based on expected energy demands. Humans do not calculate the kilowatts required to walk to the store, but they do know not to undertake certain activities because of experience with the energy demand for those activities.

All of the data sent from the midbrain to the cortex is stored in the same data planes as other sensory data. The next level up includes this data that has been spatially and temporally abstracted, so that midbrain data, such as heart or breathing rates, is captured in a single data plane beside other sensed and spatially and temporally abstracted data. Assume three prominent integrated sensations are roughly described as "wanting pleasure", "non-pain" or; "comfort" and "sense of wellbeing" and the general habit associated with these three cortically defined sensations is "happy" or "content". As long as the midbrain sensations that produce "happy" are maintained, then "happy" is included in every newly stored data plane in the input section. If that "happy" person is listening to a song, "happy" is associated with it. If a person is looking at an advertisement, "happy" is associated with it. The outcome: The person hears the song again and the "happy" data is recalled along with the stored song. The song quickly becomes a favorite. Advertisers try to create "happy" in their photo or video advertisements for the same reason. You recognize the product the next time you see it while feeling "happy" about it. There are many of these non-symbolic associations that exist because all sensed data is stored in the same data plane at the same time. You smell muffins while you are walking down the street, and you think of grandma's kitchen.

There are a few other midbrain outputs to the cortex that are important for my model. The midbrain can inhibit cortical output so that we do not move during certain sleep cycles, and the midbrain can adjudicate cortical memory-match quality and reward the cortex for better matches. The better the match, the better the reward. This is not critical for the operation of the model I describe, but it very likely exists and provides a bit more flexibility for goal-setting.

13 THE CORTEX LEARNS FROM THE MIDBRAIN

An intelligent cortex senses, through fibers, midbrain functions indirectly (fear, hunger, etc.) and the resulting midbrain instigated outputs (running, hunting, etc.) directly, and will thereby learn to emulate the midbrain functions non-symbolically (Eugene used the crying baby example). Energy management is the most obvious case. The midbrain provides energy expenditure data to the cortex, and the cortex adjusts output or activities to lessen the energy consumed, for example. This feedback loop works very well, especially for athletes that are training to increase their available energy. However, early in human history, the cortex's energy adjustment behavior became a cortex-only habit. The cortex learns to conserve energy without input from the midbrain. A human can feel tired, which is midbrain input, or they can believe that they are tired based on experiences stored in the cortex. Laziness is generally learned and, unfortunately, begets further laziness. This laziness belief is easy to ignore in healthy individuals and, with effort, can be usurped by midbrain energy expenditure data stored in the cortex.

A while back, I was hiking on the West Coast Trail on Vancouver Island with my son at an incredible pace. I was bitching and moaning non-stop about each grueling mile, the last of four days of a rather tough 46-mile hike. We passed a group of Canadians heading in the opposite direction. They were chatting and laughing and, apparently, having a good time. We stopped and talked for a moment, and I asked one of them if he was tired, even though he did not seem to be. He responded that he was just as tired as my son and I were, but he refused to act that way. "Acting tired adds no value to anything and does not change the pace or the goals," he said, so he simply stopped acting tired, regardless of how he felt. He suppressed, or never formed, the complaining habit related to when he sensed tiredness from the midbrain.

The cortex can learn the environmental prompts that cause behaviors generally invoked by the midbrain, and it can initiate different learned behaviors or outputs. The difficulty arises when the initiation of the behavior is ambiguous or in conflict with other sources. Anger can invoke a cortex-only response, midbrain-only response, or both. With sufficient midbrain stimulation either from the cortex or from the sensory fibers, it can override the cortex and produce output or actions not controlled by the cortex. Eugene used an encounter with a squirrel as an example of this behavior. There are also many newsworthy cases where relatively docile animals became enraged savages because something provoked the midbrain so strongly that their habitual response was overridden.

Generally, outputs to the motor systems transit the midbrain but are not switched or selected by the midbrain. The midbrain's programmed outputs, if any, are inhibited unless the cortex cannot provide a valid output.

As Eugene explained, cortical balks occur when there is no valid output for a set of sensory inputs. The cortex will continue to scan until it reaches an approximate match, which may not be relevant at all, or it balks and starts a new scan with newer sensory data. In adult humans, balks are rare as long as the human is in a familiar environment. In a novel environment, balks are common. As an example, I saw an eerie night-time dashcam video of many tumbleweeds blowing across a highway. In the audio, a female passenger is heard making nonsensical noises; not panic noises, not fearful noises, just nonsensical noises. She was experiencing a novel situation, and her cortex was balking. Startling or suddenly invoking fear in someone will typically cause the cortex to balk. Recovery is rapid if there is no escalation in novelty.

Military and law enforcement training is designed to expose the trainee to as many novel experiences as realistically possible so that new or old cortical habits can be associated with these experiences. Moreover, this is to ensure that cortical balks do not occur when a critical situation is presented on the battlefield or during law enforcement.

Cortical output failure in humans is even rarer. Real imminent danger is a good example of such a failure. When there are no solutions for the sensed input, the system will try to shift to midbrain or instinctive behavior. Despite popular conventions, fleeing the danger or screaming is about the best a human can do in an instinctive mode under these circumstances.

Another midbrain function is related to streams of output data planes. Walking, reciting a poem, and singing are continuous cortical outputs, which are streams of data planes. Since the data values in the data planes are discontinuous, the midbrain is responsible for smoothing the data by interpolating values between the planes. I suspect that this is the cerebellum's role.

Again, early in life, the cortex learns to mimic midbrain behaviors and is able to anticipate (because that information is in memory) the associated positive or negative results, either from the environment or directly from the midbrain, which reinforce those behaviors. Later in life, environmental responses become the sensory input which, in turn, initiates the cortical behavior that causes the appropriate, or prevents the inappropriate environmental response. The data just transits through the midbrain.

This cortex learning from midbrain behavior is not the only mechanism for developing behaviors in the cortex. It is the earliest one and very likely influences the learning of all future behaviors.

14 ENERGY MANAGEMENT

In a boxing match, each boxer is either trying to win by the judges' determination of their skills or by knocking out, or at least disabling, the opponent so the match cannot continue. Blows to the head are the most effective way to knock out or disable an opponent. Boxers parry until one finds an opening and lands such a blow. To counter the effect of such head blows, the struck fighter, if he saw the punch coming, can tense up muscles in his neck, face, and body to absorb the energy of the blow. If he did not see it coming, the full force could rotate his head fast enough to cause his brain to hit his skull, thereby knocking him out or disabling him.

(Prior to 1867 and the changes to the Marquess of Queensberry's rules mandating the use of boxing gloves, bare knuckles were the rule. Bare knuckles transmitted so much energy within such a brief period of contact that knockouts were frequent and quickly brought the match to a close. Boxing gloves, although claimed to transmit a higher-forced blow to the opponent, actually apply the impact load to the opponent's head slightly slower than bare knuckles do, allowing his muscles slightly more time to absorb the energy, delaying the end of the bout.)

Eventually, one of the boxer's head, shoulder, and face muscles would tire, becoming unable to absorb the blows sufficiently. Skillful parrying will either provide an early opportunity for a head blow or tire the boxers out until parrying becomes less effective and head blows more so.

If their boxing skills are equal, the match would be all about energy management: Hold your arms up, parry, punch, and absorb the blows longer than the other guy until he provides an opening.

Energy management is the general goal in nearly all sports and, more broadly, is the goal of all life on the planet. Most such management is provided by evolutionary changes: trees grow taller, digestive systems become more adaptable; energy storage becomes more effective. The "parry" here is to ensure that energy (food) is available when needed and current energy stores are conserved so that there is enough energy available to acquire more energy (food).

Lions rest for 23 hours each day and use the remaining time to hunt, eat, mate, and socialize. While the lion cannot worry about its next meal (it cannot think ahead to the time when it will be hungry again and whether or not food will be available), it lives its life at the ready in case a hapless gazelle wanders by. The lion lives in an environment that has sufficient gazelles to maintain the lion population, otherwise, the lion would not be there. It is not a lion's cortical thinking that manages its energy and guarantees its survival. Rather, the lion's cortex remembers the lion's territory so it does not wander too far from its community and available prey and water.

Energy management or energy balancing is an inborne function in humans. The mechanism for energy management resides in the midbrain but is eventually, at least partially, usurped by the cortex. One can imagine three gauges in the midbrain: one showing energy stores decreasing from energy being used, one showing energy stores increasing from energy being gained, mainly by eating, and the last showing energy stores. The "parry" again is to minimize energy consumption to ensure that stores are not depleted before sufficient new energy can be obtained. The midbrain alone cannot plan for gaining energy—it can only react to the environment. The cortex lets humans commit the necessary energy, based on experience, to capturing prey or other sustenance before their stores are depleted.

The midbrain manages the body's energy balance by decreasing output effort when energy sources are low and consuming excess energy for future use when supplies are plentiful. The cortex learns early on that conserving energy will gain a reward from the midbrain, as will consuming excess energy. Cortical energy-balancing habits can get out of sync with the real needs of the body and result in detrimental behaviors, such as overeating or habitual lassitude.

The midbrain has a mechanism that rewards the cortex for energy conservation at the micro-level (e.g. efficient memory matching) and provides relevant energy expenditure information (the three gauges), alongside sensory data, so that the cortex can develop a store of energy expenditure data related to behaviors. Without this data, the cortex could undertake energy-intensive tasks that are not feasible or are less efficient than others. This is an important idea. Much of our daily lives are governed by energy conservation goals. We sit to work, cross our legs, rest our chin in our hand, reluctantly accept certain work assignments, speak jargon, search for close parking places, and otherwise plan most activities to minimize energy expenditure. This does not mean that we are lazy; it just means that, good or bad, energy conservation is an ever-present part of human behavior and a specific, but partial, goal during inferencing.

The cortex itself consumes a great deal of energy under certain conditions and, like locomotion, has experience with the energy requirements for those conditions. A simple demonstration of cortical energy values is as follows: I will ask you, the reader, to commit to a number of tasks. Pay attention to your nonverbal reaction to each.

1) Memorize this sentence

2) Memorize the lyrics of a new song

3) Memorize a chapter in this book

Each subsequent task requires an increased focus or commitment of energy, known to your cortex as an approximation, before you begin the tasks. It is not a remembered word or phrase; it is non-symbolic data stored in a data plane associated with tasks similar to those above. You know (or feel) the approximate energy requirement before you undertake the task. Lions do not.

Again, I treat the midbrain as a black box and only include data paths necessary to make the cortex work. These data paths are noted on the appropriate figures. Other goals, like sex drive, hunger, and fear avoidance, are not included explicitly but clearly drive cortical behavior. Hormones also affect the midbrain, and the midbrain provides related data to the cortex which, in turn, will provide a learned course of action.

I am avoiding great complexities in explaining the cortex's role in human behavior by eliminating hormonal- and midbrain-generated sensations. This section further reduces the problems by elucidating the differences between midbrain and cortex energy management. Energy commitment data is stored within every stored output data plane, thus providing a goal or measure of viability for any selected output.

Since the energy expenditure data is provided by sensors, such as baroreceptors, temperatures, and heart rate, it too must be and is temporally abstracted to provide areas in data planes related to total energy expenditure that are, in turn, related to behaviors also represented non-symbolically in the same data plane. Knowing a person has enough energy to outrun a predator or to swim across a lake was hugely beneficial to early man.

With this said, I treat the midbrain as a source of information for the cortex about the body and, to some extent, a purveyor of appropriate cortical output to the body.

15 MEASURING INTELLIGENCE

Intelligence is perceived in many ways. A person stating many things that others accept as facts appears intelligent to outsiders, even if they are simply reciting from memory or rote and cannot relate the facts to other environmental conditions. The more facts you know, the smarter you appear. (There are many cases in which the facts are not valid, but the listener is unaware of this, so the speaker still appears intelligent.) This is iterative intelligence and makes up most of the intelligent behavior communicated amongst humans. Eugene postulated that about half of cortical energy expenditure is used for iteration and the other half for association. This is his theoretical balance. In humans, this balance does not exist.

A person using perceived facts (e.g. someone told them something) can relate those facts to other environmental conditions stored in memory through inferencing and produce or postulate new facts. This is associative intelligence.

The cortex can accept anything as fact (i.e. stored in memory) as long as the output associated with these facts produces a desired (rewarded) result. This is problematic when considering intelligent behavior in humans for three reasons: The invalid output can be inscribed in the environment for others to absorb as fact, the holders of these pseudo-facts can successfully infer other invalid facts that can be inscribed in the environment unless feedback from the environment caused the pseudo-facts to fade from memory, and the holder of that information can continue along the same path of beliefs without making any adjustments. This is not intelligence, although many people will argue otherwise.

Science is a community of people inscribing valid facts in the environment to provide input data that will allow intelligent behavior in others. This is the evolutionary advantage that ensured the survival of humans in more recent times. The collective community becomes smarter and more capable of adapting to a changing environment because a broad store of valid information is available to us. These inscriptions began as cave paintings and petroglyphs. These inscriptions were not decorations; they were encyclopedias for the community. Spoken words were an early source of inscribed knowledge but were limited to a local audience and by the frailties of poor recall and misunderstandings. Only when written inscriptions were developed did knowledge transfer become consistent and far-reaching.

Some people seem to thrive in the iterative mode: talking, reading, and studying the environment. Levels of valid data in iterative conversations or writings vary widely, making an assessment of iterative intelligence difficult. It is straightforward to measure the facts that someone knows, as most testing in schools does, but it is far more difficult to measure how well information is stored and recalled. A neurological problem with memory, spatial abstraction, and temporal abstraction or a learning void look much the same in test results. Early learning voids are very difficult to overcome in later life.

Other people seem to thrive in the associative mode, thinking and daydreaming, but with less validation interaction with the environment, especially through conversation. This can lead to a buildup of stored but non-validated inferential results. Done over time and to excess, behavior considered as acceptable in the environment suffers as it is supplanted by incompatible inference results. Psychiatrists and psychologists spend a great deal of time trying to unwind this inferred but invalid knowledge to get a patient's behavior better aligned with environmental expectations. To be accurate, they also spend a great deal of time trying to unwind bad iterative knowledge. These are two very different mental issues.

Measuring associative intelligence is difficult. Eloquence in writing is a good measure of symbolic inferencing. Non-symbolic problem solving, or figuring things out, is a good measure of non-symbolic inferencing. Certain Mensa tests are great examples of non-symbolic problem solving because they do not depend on a store of facts or the understanding of symbolic language. However, it is still difficult to determine whether a problem-solving test result indicates a neurological deficit or a learning deficit. There are many mechanisms used in inferencing that can produce exceptionally good or bad results. Getting high scores on associative tests depends on the correct developmental conditions, some physiological and some informational. Deficits here are also very difficult to overcome.

Standardized IQ test results generally have a normal distribution. 1 in 44 people achieve a score of about IQ130 which is consistent with a normal distribution. However, 1 in 100,000 people achieve an IQ190 score, when the normal distribution indicates that only 1 in 2,000,000 should. There is no law that says you can't have a lot of really smart people but it is very unusual that 20 times as many people could correctly answer what has to be the toughest questions in the tests than expected. This is such a large deviation that it is very likely caused by a physiological difference and not a learning difference. I suspect that cortical enhancements (deeper memory, broader specialized areas at the expense of other specialized area, bigger or different fiber tracts for inferencing) evolved more recently and are separate linages, surviving alongside the common linage, in just a few individuals. Those linages will either persist, merge and become part of the common linage, or die out, depending on environmental pressures. Another less likely option is the test questions. The most difficult questions that require memorized information including a procedure to solve, are so difficult that very few iterative thinkers can answer them. Unexpectedly, the inferential thinker at the high end of the scale figure out, or infer sufficient procedures and information to answer those questions correctly resulting in an unexpected higher score.

With a structure like I describe herein, there are new opportunities to define and measure intelligence.

Figure 4 shows a timeline for specialized area development and a rough illustration of rich and poor development. Poor early development yields poor later development, unless significant effort is made to offset the deficit.

There are several measurable functions that relate to intelligence:

Four things that drive learning

1) Physiological completeness and deficits, including

    a. Genetic physiological differences

    b. Differences caused by DNA errors

    c. Disease and injury

    d. An excessive imbalance between cortical and midbrain authority

2) Early learning, including very early sensing and SA, TA, and fiber tract development at lower layers caused by either

environmental stimuli or a lack thereof, a lack of energy, or reliance on midbrain inputs to the cortex

3) Mid-level learning, including upper-layer SA, TA, and fiber tract development caused by a lack of feedback from the environment, a lack of energy, a lack of early learning, or reliance on midbrain inputs to the cortex

4) Diminished inferencing due to habit that can be caused by either a lack of successful inferences or a lack of energy

The ability to pay attention is critical to developing intelligence. Paying attention, or the ability to focus, is a key behavior that is poorly understood. Eugene wrote about up-pacing memory scanning, when interesting things are sensed in the environment, and free scanning, when input is more common.

Neither sensory input nor memory scanning can be turned off. (Sleep is a special case.) I doubt whether memory scanning speed can be changed, so Eugene's up-pacing would have to be inferencing or daydreaming, when input is dull, and iteration, when input is novel. Signal propagation speed down an axon cannot be changed and, if it could, would create an intractable synchronization problem within data paths.

This probably means that sensory input has priority for energy (blood flow to the appropriate lobe). With low match rates, the midbrain can shift blood flow to other lobes, particularly the temporal or frontal lobes, facilitating inferencing. Non-symbolic inferencing probably has priority over symbolic inferencing because sensory input is non-symbolic and real-time. Regardless, this focus-shifting action is measurable and learnable.

The low-blood-flow lobes are still working, but the memory-matching quality is poor. Any newly stored data is invalid and will be purged during sleep cycles.

This is one of two mechanisms that prevents locking into a non-varying loop of input and output or catatonia. The memory-match reward stimulus from the midbrain is the other mechanism. This stimulus dwindles from overuse and memory-matching quality decreases, resulting in partial or inexact matching in other parts of memory.

There is a clear evolutionary advantage to sensory lobe priority. If this priority is genetic, it must be accounted for in any testing standard.

A fully developed brain should be able to match and partially match, within a prescribed range, a variety of inputs and produce valid measurable outputs.

Iteration is measurable, as are symbolic and non-symbolic inferencing. The depth of memory and the effectiveness of each inferencing loop are clear indicators of intelligence. How much content is the result of symbolic inferencing with strong matching and non-symbolic associative inferencing which, in this case, is inexact but useful matching? How much of this symbolic inferencing data buildup is the result of interacting with the environment versus being internally generated?

Inferencing is behavior and, as such, can be suppressed or enhanced with environmental feedback. Is poor inferencing behavior suppressed or was it never developed?

Inferencing can occur within a cortex, both symbolically and non-symbolically, although symbolic iteration and non-symbolic inferencing are prevalent. Iteration also occurs between the cortex and the environment, both symbolically and non-symbolically. Furthermore, iteration can occur between two cortexes and even some variations of the cortex and midbrain. These combinations will change as the experience of the cortex increases but should be accounted for in cortex development and operation during any attempt to measure intelligence. Another important distinction is determining the overall goal of measuring intelligence. Is intelligence measured against the tested abilities of the brain versus its specific measured capacity or is it tested against an external standard?

The standard for intelligence is the characterization of a population using various tests. A better standard might be to test whatever best suits the human in his or her environment. Better still would be the quantification of the characteristics above to form a cortical characterization statement that could be compared to others in the same environment. As a teaching aid, the cortical statement could be used to customize learning plans to overcome early deficits.

The statement might look something like this:

Data ledger content

    Iteration counts over time

    Symbolic inferencing counts over time

    Non-symbolic inferencing counts over time

And perhaps

Iterative error rates

Symbolic inferencing error rates

Non-symbolic inferencing memory-matching range

The correlation between symbolic and non-symbolic data (the degree of integration between the two). Do they integrate correctly at the top layer?

Discontinuities between the top-level data planes

The width and depth of memory planes and specialized areas

And, more critically, the length of symbols (focus epoch)

Since iteration and inferencing can produce the same results (i.e. if one fails, the other may produce valid results), the characterization statement could reveal deficits not detectable through current testing methods.

The characterization statement reveals nothing about the depth and breadth of information content, but it would be a good indicator of a person's intelligence capability. As with Eugene's data ledger, this cortical characterization statement would be very difficult to measure.

16 CORTICAL INTELLIGENCE THROUGHOUT HISTORY

As I searched for historical indicators of evolutionary cortical changes that would account for various aspects of intelligent behavior over the past 100,000 years, several things became apparent. Evolution did not and does not occur as it seems or as we have been taught, and there is not a single evolutionary path for humans, intelligent or otherwise. The current scientific categorization of evolutions is represented as a tree diagram, with each branch indicating a new species, and is known as a phylogenic tree. Archeologists and other scientists do an amazing job of defining the branches of the tree with scant evidence. The differences among branches on the phylogenic tree appear great because archeological data indicates that there are large variations in the creatures that make up each branch. In my view, these gaps should be narrower or, at least, narrow at the bifurcations or branch points. There are few evolutionary reasons as to why we do not or did not have tails, for example. Why do some humans or other living things not have the features that would bridge those gaps? To further add to the confusion, there are significant evolutionary undershoots and overshoots: features that should not have gone away and new features that evolved but were not needed. Due to the latter, today's humans exist.

As a phylogenic tree would suggest, about 3 million years ago, a female *Homo Erectus* had a genetically different female baby, and that baby was designated *Homo Heidelgensis* in modern times. That baby matured and somehow proliferated her genetic changes and, a few hundred thousand years later, a descendant female *Homo Heidelgensis* gave birth to a *Homo Neanderthalensis*. A relatively short time later, another *Homo Heidelgensis* female gave birth to a *Homo Sapiens*. This branching of the phylogenic tree is, of course, an artifact of the representations and simplifications in the writings on evolution. There never was a single female that gave birth to a new species.

Humans and, in fact, all living things are defined by a few dozen very long DNA molecules. Each molecule consists of two very long strands connected by links of four specific nucleobases, forming a ladder-like structure that twists into a helix. The sequence of ladder rungs, or nucleobases, is a code for producing the various proteins necessary for both the development and sustainment of living things from the original egg cell. During reproduction, male and female DNA molecules split down the middle. A female half and a male half combine in the egg during fertilization. The alignment of the two halves determines the genetic makeup of the offspring. This alignment is the evolution mechanism.

Evolution is discrete, but only at the genetic level. An expression of a new protein caused by the inclusion or exclusion of different nucleotide base pairs in a cell's DNA or the doubling of an existing protein nucleotide sequence are examples of possible discrete genetic changes. Differences in human skin color, hair color, height, intelligence, etc. appear to be continuous, but they are, in fact, discrete increments. Iris color is a clear example: There are no red, purple, or yellow irises. The proteins that provide the iris' color attribute are coded from discrete genes within a strand of DNA. Other variations in these genes do not work for whatever reason, so iris color is limited to the colors provided by the genes that do work. The mapping from genes to human features is not one-to-one. The code for a liver enzyme protein might also cause a different iris color and, perhaps, a slightly improved metabolic mechanism.

There may be variation in attributes generally caused by the environment; height and weight for example. However, a genetic change will be a discrete difference in one or more attributes, with the ancestral attribute having a different but likely overlapping distribution. Dwarfism is an example of a persistent lineage. There is a large population of humans whose heights fall within a normal distribution. There is also a smaller but still large population of people with the genetic dwarfism difference whose height falls within a different but overlapping normal distribution. There is a discrete genetic change that allows these two height distributions to exist. Is there nothing between these distributions? Maybe. There could be other genetic changes that affect height but little else, so they are not recognized as being out of the norm but are actually a discrete genetic change. It is this genetic discreteness that causes the phylogenic tree's branching, even though only some of the branches continue on to become a separate species. The gaps between the branches are filled in to a certain extent, as limited by genetic discreteness, but the branches themselves are actually many lineages, especially at the initial branching. Eventually, a few lineages become dominant within an environment and the rest die out. The few, then, become a fairly specific branch of the phylogenic tree. The process then repeats itself as environmental demands change.

A lineage branch is formed with every genetic change. Most members of a branch do not gain or lose much when compared to the ancestral branch or even current adjacent branches. A few will be at a disadvantage and die out and a few will have advantages that allow them to proliferate better than members of other lineages but, for the most part, the remaining lineages breed with other lineages so their distinctive traits become dominant or recessive in a common lineage. Eventually, the successful lineages will lose their ability to breed outside the lineage and become a species. There is no need for a single female to give rise to a new species, so there are no issues regarding having compatible male mates. The females in a community will give birth to babies, all of which are genetically varied. These babies will grow up and mate with other members of the community and produce offspring that have different lineages. Some lineages will provide advantages in the environment and proliferate, while other lineages will die out or merge with other lineages. As a lineage's population grows based on its DNA advantages, a DNA change would eventually occur in the reproduction mechanisms that would prevent breeding with other less successful lineages, and a new species would be formed. The rest of the original lineage carrying the advantageous DNA would eventually die out through interbreeding and diluting the advantageous genes. A community of creatures simply becomes a new species over time.

Another important misconception about evolution is that only changes that adapted an organism to its current environment survived. In fact, evolutionary changes are fairly random, so there must be many changes that are not needed to survive and proliferate that prevail. I suspect that evolution is not entirely random and that the information in the non-coding portion of DNA somehow prevents certain previously tried sequences from being incorporated into the next generation's coding sections. Moving a gene from coding to non-coding would have to happen at random as new lineages are formed. A gene that prevents proliferation could obviously not be moved to non-coding without creating a non-viable lineage, so random changes, good or bad (but not lineage-ending), would be captured in the non-coding DNA without harming successful evolution. The prevention of necessary gene proliferation at random would end the lineage and eliminate that set of non-coding instructions so no harm to the original lineage is done (because other members of that linage would continue to proliferate). This is a powerful evolutionary mechanism. Necessary genes are preserved; genes that are not necessary are eliminated from the coding sections, which results in lineages with some advantages or, at least, no disadvantages. If a gene that is locked into the non-coding section is needed again, it would re-evolve. This prevents evolution from moving backward or moving at random and gives the appearance of intelligent guidance or, at least, planned progression.

The argument here is that evolution does not select the fittest. Anything that can get by will get by. There are many evolutionary overshoots and undershoots that do not seem to affect survival. In fact, all genetic changes are overshoots or undershoots and every new human is genetically different.

The evolution of life on earth has been driven by energy management; organisms must adapt to their food or fuel supply. For most of evolutionary history, this energy management balance was shaky at best. The organism acquired just enough fuel to survive and proliferate. As organisms developed better brains and mobility, energy management became less risky, and the surplus energy or surplus time could be used for grooming, socializing, and exercising excess brain capacity.

One evolutionary pathway to improve energy management was to improve mobility, expanding the range of hunting and gathering as well as escaping danger. Four-legged creatures began walking upright, which required a lighter upper body. As the upper-body size decreased, lung size and fat storage decreased, so buoyancy decreased, and pre-primates lost their inborn ability to swim. Apparently, the birth rate has been high enough to offset the drowning rate since then, so this evolutionary undershoot persists today.

Another evolutionary pathway to improve energy management was a bigger brain. The discreteness requirement resulted in a substantial overshoot in energy management and, later, an associated overshoot in cortical evolution. We have substantially more brain capacity than we need to survive in the environment that originally produced us. The human ability to relate sensed things across time allowed us to live in large groups, defend those groups, develop agriculture, and thrive, while earlier pre-human groups or even earlier human groups could not compete with these abilities and died out.

Human music skills, oratory skills, fine art skills, etc. are examples of human behaviors made possible by cortical evolutionary overshoots that are an amalgamation of other random evolutionary changes and are not needed to survive and proliferate. The cortical overshoot is not specific to these skills. These skills developed because the excess cortex capacity was available.

About 10,000 years ago, intelligence as we define it today emerged in humans; humans with smaller brains than their ancestors. The mechanism for this had to be plausible within the limitation of the physiology at the time, had to comprise only a few discrete evolutionary steps, and had to account for the smaller brain size. Spoken communications within human, primate, and other species had existed in a rudimentary form for hundreds of thousands of years. Most were guttural utterances that embodied observations in the current environment; a warning about predators in the area, for example. In pre-humans, the utterances became more diverse but were limited by face and jaw shape, as well as by limitations in the vocal tract itself. These limitations were overcome in various human lineages as evolutionary overshoots that eventually allowed the smaller-brained humans to develop complex languages. The intelligent mechanism was not caused by the emergence of languages; intelligence allowed for the development of languages due to the new cortical capabilities.

Evolution comprises two steps: First, new genetic variations occur in a community. Some of these variations provide advantages and, therefore, form lineages. Second, something happens to prevent the genetic dilution of these lineages that would cause the loss of the advantage. One such protection would be a social adaptation of only breeding within a lineage.

The local geography also protected lineages from dilution. Mountain ranges, oceans, deserts, rivers, and other such features would contain the propagation of the genetic advantages of a lineage within geographical regions. Dilution is still possible, since genetic speciation may not occur fast enough to prevent it. As an example, I suspect but cannot confirm that the flightless cormorant of the Galapagos Islands can successfully breed with their flying kin in North America. Their geographical separation may have only created a lineage and not a new species.

This discussion has little bearing on the more modern and subjective ideas related to bloodlines, bloodline superiority, and bloodline dilution. In this writing, lineage is a genetic construct.

Neanderthals lived in Eurasia from about 200,000 years ago. They had no language but used tools and lived in communities. They made clothing and survived in the harsh climates of Europe and Central Asia. Cro-Magnon, a *Homo Sapiens* lineage, appeared about 40,000 years ago, had a language, and apparently had an easier time adapting to their environment. Neanderthals learned by observing the behavior of other Neanderthals. Learning the collective knowledge of all nearby, or more distant, Neanderthals was limited. Except as noted below, they could get through winters as long as it did not require experiences not observed in the past. The Cro-Magnon's language allowed them to capture past experiences and convey them to others, giving them a significant advantage over Neanderthals. The ideas conveyed by the language embodied regional and temporal knowledge not observable by the Neanderthals. Cro-Magnons adapted and survived because of this. This was the beginning of symbolic iteration, first with other Cro-Magnons and then, probably, internally.

It has recently been discovered that modern humans of European or Asian origin have between 1% and 4% Neanderthal DNA. It seems the genetic change to prevent cross-lineage breeding was not completely successful when the *Homo Sapiens* branch began. It also appears that the Neanderthals passed only some very important DNA changes to *Homo Sapiens*, probably from the Cro-Magnon lineage. Common interbreeding would have diluted the DNA lineage to far lower than 1%. So, for more than 1% to prevail, these DNA changes must have provided an advantage.

What is in the 1% to 4% that has been preserved? This is an interesting scenario that illustrates my point about lineages and advantageous DNA changes. Unfortunately, there is no valid evidence for it today. The Neanderthals reached the Mideast and Europe 50,000 years ahead of *Homo Sapiens* (Cro-Magnon). As they migrated northward, they encountered climate changes, for which they had no genetic adaptation. They struggled with the harsh winters, especially staying warm and finding food. They lacked a language, so there was no way to preserve information about surviving the winter beyond direct observation and, even then, it was unlikely that Neanderthals knew that weather was cyclic. Sometimes it just got cold. When it got cold, members of the community died and most went hungry. A simple genetic change could have caused the Neanderthals' feeding habits to change and could account for their survival. The hormonal and neurotransmitter controls for hunger and satiety are different for fats, proteins, and carbohydrates. A genetic mutation, in which Neanderthals craved carbohydrates or were not satiated until carbohydrates were consumed, would cause them to gain body fats in the fall, when various wild vegetables, nuts, and fruits ripened and were available. This body fat would have provided them with the energy to get through the winter until hunting once again became possible. In this case, the midbrain adapted to climate change because the cortex could not relate to the various conditions nor provide a solution. (Today, in humans, carbohydrates metabolize about 30% faster than fats or proteins, so consuming carbohydrates will tend to increase body fat if they are consumed at the same rate or volume as fats and proteins. High-protein diets and, particularly, ketogenic diets leverage this evolutionary artifact to effectively reduce stored fat.)

Now, imagine a community of Cro-Magnons, a lineage of *Homo Sapiens*, migrating north towards Europe 20,000 years later, from warmer climates to the south. As winters approached and the weather got progressively colder, game became scarce, as did grubs and tubers. The hunter-gatherers of the community continued their hunt as usual but, for each foray, they returned to the camp with less and less food. There were no discussions among them about better hunting tactics or strategies for staying warm because Cro-Magnons had no experience living in cold climates. Villagers just got cold and hungry which, as with the Neanderthals, caused some to die.

At a fork in a large river, too large to cross, Cro-Magnons gathered and camped, as did Neanderthals and maybe other lineages. The

Cro-Magnons eventually, directly or indirectly, killed off the Neanderthals but not before interbreeding with enough of them to supplant the Cro-Magnon lineages with a new, yet unnamed, lineage. The members of the new lineage had a genetic attribute contributed by the Neanderthals that made them crave sweet fruits and tubers.

Life in the spring and summer for the new lineage was normal but, in the fall, when the fruits and tubers ripened, they gorged themselves until the food was gone and they were fattened. Then, in the winter, food again became scarce, but more of the new lineage survived because they carried enough body fat to get them through the winter. The Cro-Magnon had skipped 50,000 years of evolutions and certain extinction by exchanging DNA that had already evolved with the Neanderthals to survive the cold winters and ice ages. There is non-coding Neanderthal DNA in a majority of all non-African humans today, which means at least some part of this scenario is true and most, if not all, other non-African *Homosapiens* lineages died out in the past two ice ages. The idea of transferring the sweet-craving trait from Neanderthals is interesting but has yet to be borne out by scientific studies. However, the sweet-craving genetic trait is real.

Pre-human and human migrations tend to sound like planned acts but, in fact, if food became scarce or living conditions too dangerous, people just moved somewhere where there were fewer people. Eventually, they moved from Africa into Central Asia and Europe. They also migrated across Asia to eventually inhabit Indonesia and the Pacific Islands. The migrations to the islands are far more recent, since boat building and sailing only appeared within the past 5,000 years.

(A friend of mine once marveled at the insight of the ancestral Polynesians that sailed to and settled in Hawaii. I ruefully pointed out that the Polynesians had no prior knowledge that the islands were there and had just set sail from Tonga in all directions and, over time, a few had landed on the islands that then came to be known as Hawaii.)

Once in Europe and Central Asia, when food became scarce or living conditions dangerous, the ability to successfully move somewhere else was diminished because these areas were bordered by mountains, deserts, and oceans. These geographical features created population congestion not experienced by humans in the past. Eventually, to survive, growing food and defending families from marauders had to become a collective effort. Organized groups or kingdoms, much larger than family- and village-sized units, afforded better survival, but that organizational construct stalled until about 10,000 years ago. Around this time, the cranial capacity of *Homo Sapiens* began to decrease, working its way down from about 1,500 cc to the current 1,200 cc average volume. There are many discussions regarding the metabolic energy commitment required to sustain the larger brain, and the decrease in female pelvis size that limited the successful birth of larger-brained lineages was prophesied as the cause of this decrease. The bigger-brained *Homo Sapiens* existed but needed a smaller brain to get by? This seems nonsensical. It is unlikely that intelligence was sacrificed for a smaller brain volume. Obviously, intelligence increased.

(Some interesting conjecture here, is that as analog signals propagate along axons that make up a pathway, the signal diminishes. In larger brain human and animals with the attendant longer pathways, temporal abstraction didn't work because the signals were too weak at the distal end of the fibers to accumulate to within any workable scale range. Early big brained humans lived in a world of iteration and could not summarize or abstract anything. They could match input to memory and produce a learned output. (Idiot Savant is a contemporary description of a physical condition that precludes abstract thinking. The physical condition includes changes to the surface topology of the cortex and thus fiber lengths.) Apparently, there wasn't an evolutionary path to increase the signal strength but there was one to shorten the fiber pathways, a smaller brain. As the brain got smaller, temporal abstraction began to work and the substantial changes in human behavior began.)

This change is problematic in that this decrease in brain size must be of evolutionary origin. It apparently occurred 10,000 years ago and originated in Europe, yet the smaller brain size is seen throughout the *Homo Sapiens* population. There is currently only an approximate 100 cc variation in brain volume across all studied *Homo Sapiens* lineages. The cranial volume decrease must have occurred at a common human lineage branch of the phylogenic tree, discovered or not. This lineage survived alongside or within the common lineages for 40,000 years, about the time the *Homo Sapiens* lineages were still centralized enough to ensure an even distribution of the trait and only emerged 10,000 years ago as a dominant lineage because the other lineages could not survive in the geographical and relatively crowded confines of Europe and the Middle East without the increased ability of the smaller brain. Since the small-brain lineage emerged before humans left Africa, it eventually prevailed in some form in all humans.

The temporal abstraction and cortical inferencing mechanisms would have solved the problem of a needed increase in intelligence with a smaller brain and must have evolved into a useful cortical function at about that time. The lineages of humans without the cortical abstraction trait would not have been able to compete with humans that could relate things across time or related groups of things into single entities. They would eventually have died out, leaving no archeological trace of their cortical deficit.

The larger ancestral brains were needed for remembering territory etc.: the larger the brain, the larger the territory or the larger the understanding of food or other creatures in the environment. *Homo Sapiens'* smaller brains could abstract and otherwise infer results and, therefore, did not need this extra capacity.

The temporal abstraction of language (symbols) only produces more language (higher-level symbols) and not non-symbolic representation of sensed information. Symbolic data must also be a source, like sensory sources, for integration in the non-symbolic cortex. It must be spatially and temporally abstracted, along with other relevant sensory data and experience. Once this mechanism was in place, non-symbolic reasoning about things learned from the language, both iteratively and inferentially, was possible. This connection between non-symbolic and symbolic reasoning was probably made or completed about 10,000 years ago and accounts for the human ability to form and act as part of a larger, but unseen, single entity, generally thought of as a government or organized religion. Eugene suggested that the insular lobe, which is small, is this linkage, and I suspected that it was the most likely source of this evolutionary leap.

The abstraction of many similar things, originally conveyed symbolically, to a single entity provided the mechanism that allowed humans to behave collectively. There could then be providers and protectors, nearby or far away, that, as a single entity, had abilities and characteristics of its own. Humans could infer or recall benefits attributed to the single entity and take courses of action that would let them or their family benefit from relationships with the single entity, even though these relationships were manifested through individuals.

This abstraction ability applies to everything: the herd (of cows), the government, the children, the crop, the fertile valley, the enemy, etc. Each of these single entities has its own characteristics that may not apply to all members or, perhaps, any member of the abstracted set. These single entities could and do have many attributes associated with them, obtained through the course of human learning.

This abstraction ability also served the individual's own perceptions, which led to the non-symbolic abstraction called self-awareness or the "me habit." This habit must be embedded in all top-level non-symbolic data planes, which is why it is so difficult to overcome.

Even though we like to think that the family pet can behave intelligently, I doubt whether most animals have any abstraction ability. (As Eugene explained, we have a strong bias to over-animate things, especially our pets, with our thinking abilities.) As an example, a dog might sense an item in its environment as just that thing. If the thing is divided into three pieces, it is no longer the original thing; there are just more different, individual sensed things. The dog may be able to recall the original thing if it sees it again, but it cannot relate that thing to the three new things.

A language (symbolic) cortical mechanism gave the Cro-Magnon a survival advantage. Crowded conditions further advantaged integrating language (symbolic data) with non-symbolic data, allowing the abstraction of ideas and things, which then allowed a distribution of resources that benefited the larger population. Evolutionary needs drove the rapid extensions of cortical functions. The environmental successes of the Cro-Magnon, provided by the increasing numbers of abstraction levels, were driven by evolutionary change in the connectivity and structure of the cortex, more than likely in the frontal lobes. There was not much genetic change required to manifest such capacity increases.

The evolution of data processing in the cortex roughly occurred in this sequence:

Spatial abstraction is very primitive. It must have existed in every animal with sensors and memory abilities.

Non-symbolic iteration is also primitive and likely first occurred with spatial abstraction. It allowed the creature to learn from its mistakes, without a need for language or instructions.

Temporal abstraction may be specific to humans. It allows for data integration across time and, therefore, the abstraction of information. This is a single-entity mechanism, but true abstraction requires inferencing.

Symbolic iteration and inferencing are required for speech. Uttering something and getting a useful response is probably reserved for humans but not the earliest humans. This a significant step in intelligent behavior and probably emerged 10,000 to 20,000 years ago.

Non-symbolic inferencing is required for associative thinking. It most likely appeared around the same time as symbolic iteration and inferencing. Inferencing, or figuring things out, greatly reduces memory capacity requirements. The advent of all inferencing is very likely associated with the decrease in cranial capacity. Associative inferencing, especially procedures, requires unique dedicated pathways that evolved very recently, first in lineages then as a trait, as the lineages merged into the main lineage. Indeed, kings, nobility, and clerical leaders throughout history may actually have risen to their positions of authority, not because of their education or station in life, but because their linage included the earliest inferential pathway trait (they were very smart). Nobel bloodlines may have actually been inferential pathway linages. The Nobel's could figure out how to avoid being overthrown and how to maintain an independent linage better than others.

The integration of non-symbolic and symbolic inferencing is required for executing procedures, planning, and true abstraction. It is probably associated with the improved function or capacity of the insular lobe or another mechanism that mediates communications between the symbolic and non-symbolic sections. This is the most recent upgrade to the intelligent mechanism and is less than 10,000 years old. This inferencing ability also allowed for the association of attributes to single entities.

Some of these are conjectures based on the idea that brain function changed incrementally over an extended period. Some of that time was used for waiting for the next increment of evolution, and some was used for waiting for the environment to develop sufficient speech, for example, that could allow for speech abstractions.

An interesting anomaly in the history of the cortex is music. There is no identifiable evolutionary advantage to being able to process instrumental music, let alone sung music. (Granted, some birds use instrumental music for limited communications.) Historically, grunts and chants of early man are sometimes considered a form of music that dates back about 50,000 years. These grunts and chants are not aligned with the academic definition of music, with prescribed pitches, scales, and tempo, which appeared about the same time that I claim the top-level non-symbolic and symbolic iteration and inferencing appeared. Music is simply an environmental exploitation of a cortical capability. The brain can do it, even though it was not necessary for the evolution of the brain. This category of exploits would also include fine arts and oratory skills. I suspect that there is a strong correlation between the structure of instrumental music and the structure of the cortex. For example, metric or rhythmic range may be proportional to the temporal abstraction epoch in the first stage or memory scanning speed. It is safe to say that music and the fine arts exploit the capabilities of the cortex and are a likely source of measures for these capabilities. (Instrumental music is non-symbolic, so there is a chance that there was an undetected evolutionary advantage for communications with melodies, harmonies, and chords. The ability to communicate using symbols and language required a large number of evolutionary changes that allowed for larger and larger vocabularies, but this took a great deal of time. Communication between early humans before and during that period could certainly have been supplemented with whistles and tones in complex sequences that described environmental benefits and dangers. Certainly, geographical information could be conveyed in this way, similar to how whales communicate do today.)

Evolution has not stopped, nor will it, ever. We are tall and short, light and dark, sensing-dominant or inferential-dominant; these are genetic traits amongst thousands of others. We are all different. The environment is constantly changing, so the traits that will ensure our survival will also shift into dominance. Human intelligence traits do not have a guaranteed position on that list.

In fact, the environment will very likely change enough in the next few hundred or thousand years for a future lineage with certain advantages to eventually become a successor to the human species. Perhaps intelligence, or even extensive memory, will not be needed, as machines will provide such services. Automated low-impact food growth and preparation may even provide everyone with sustenance at no cost or effort, obviating the need for humans to work or shop for groceries. Mobility will also become unnecessary. Regardless, increased cortical intelligence will not evolve and prevail unless there is a place for it in the assurance of the survival of the human species.

17 BEHAVIOR

There are four or five (or six) kinds of taste buds on the human tongue that account for all of the flavors perceived by humans; around 10,000 individual tastes or taste combinations. A typical individual can also identify many odors. The sensory data from about 12,000,000 olfactory sensors and 10,000 taste buds are first routed through the midbrain, where the data is used to avoid bad food, eat mediocre food, and perhaps gorge on good food. This data is then consolidated and presented to the cortex as input. This involves around 1,500,000 input fibers. Another 1,000,000 fibers carry visual data from each eye, 500,000 fibers carry proprioception data from the body, and another 50,000 fibers carry auditory and vestibular data from each ear. Auditory and visual data contain symbols that are separated into symbolic areas on the cortex. All of the non-symbolic data are integrated into non-symbolic areas on the cortex after being sampled by the midbrain. Taste, smell, non-symbolic sights and sounds, body motor status, and body health status are included in this category. Body health includes proprioceptions and hormonal messengers conveying hunger, thirst, sex drive, irritability, pain and pleasure, energy output, and certain sensed social precursors, such as family smells. The cortex records this data in memory, spatially and temporally abstracts various input tracts and, either through experience (iteration with the environment), correlates it with earlier, previously experienced data or finds relationships with earlier, unrelated experience data (associative inferencing). Either way, results, as consequences associated with the stored input, are executed as output to the body. Even with this vast sensing and correlating system, physiology still limits the human cortex and, therefore, human behavior to focusing on one or two subject matters at a time. More extensive correlations are accomplished by inscribing various conclusions in the environment and having humans correlate a couple of them and then re-inscribe the results. A human simply cannot correlate a dozen different input streams or subjects, iteratively or inferentially. Priority of focus is generally driven by pain or pleasure or, more likely, the learned behavior of "avoid pain and seek pleasure". Both sensations are broadened to include other experiences (i.e. The habit of seeking pleasure is associated with other things. Happiness is a learned variation of seeking pleasure.), so a sip of excellent wine is more pleasurable than a sip of good wine. Minimizing energy expenditure or creating a plan or procedure that allows for minimum energy expenditure is a common goal whose result is rewarded after the fact by recording the energy expenditure of the successful planned endeavor. Failure of a plan due to excess energy requirements is recorded as such and not repeated or retrieved again as a viable plan. Lastly, memory matching is rewarded physiologically with, for example, a little "squirt" of pleasure from the midbrain for every good to excellent match. As above, the behavior learned from this reward sensation is broadened, so that making good to excellent memory matches also becomes habitual.

As symbolic things are sensed, they are spatially and temporally abstracted and then stored at the next layer of memory in the cortex. This can continue until the information reaches the top layer of integration, where it becomes usable for constructing other symbolic outputs. You can learn a new word or read a new sentence or book, and that information will be stored in certain top-layer data planes. You can then use the summary level, or any memorable lower level, to describe what you read. You may not understand some or even all of it! You can use words that you cannot attribute meaning to correctly or, at least, nearly correctly in a sentence. Advertisers (and politicians) count on this oddity to keep their brands or names fresh in your memory without you having a reason to do so.

Understanding something means that it has been integrated at the highest non-symbolic layer of memory. It is included in a set of data planes that also include other things that have been related through inferencing. "I understand" should be a sufficient indicator of establishing these data planes, but "I understand" can also just be a habitual response to a question. Most humans, however, know if this non-symbolic relationship has been established. "I get it" is a common utterance indicating this understanding. Most humans also know when they do not "get it." For instance, one can pass a calculus course but not understand the purpose of calculus nor how to recognize when to apply it.

Again, Eugene postulated that the cortical data ledger should have about 50% iterative data and 50% associative data as an estimate. This would mean that a human would learn about 50% of all learned things from the environment and would figure out, through inferencing, the other 50%. Without the first 50%, nothing can be figured out correctly. Some people live almost entirely in the iterative world, relying on everything they hear as being true statements that can be re-iterated. Through early learning shortcomings and bad thinking habits, their inferential mechanism or habit is poorly developed and untrustworthy. On the other hand, some people rely almost solely on inferential thinking to the detriment of conversation and iteration. If the underlying store of non-symbolic data is replete with valid information, this thinking style works well, although a shift back to iteration to interact with the environment is sometimes clunky. Unfortunately, inferential thinkers can lack valid data. Because associative matches can be inexact, they can produce information that sounds good but cannot be traced back to substantive facts, stored or otherwise.

Most personality tests correctly categorize iterators and inferentials, although their ratings are typically tainted with levels of energy and focus.

There are a few mental hazards associated with symbolic iterating versus non-symbolic inferencing. The highest level of behavior is stored in the highest and most abstract levels of non-symbolic data planes; one can talk without the symbolic section interacting with the non-symbolic high-level data planes. In some cases, prohibited or less-than-optimal behavior may not be inhibited. In other cases, the syncing of the talking side with the non-symbolic side may be insufficient and result in separate and different allowable behaviors. Generally, this problem manifests itself as someone saying something, even though they know that they do not believe it or that it is not true, or they have no valid information about the subject but talk about it anyway.

The reverse presents a different set of behaviors. A person can thrive in the non-symbolic, or daydreaming, section and can construct an entire set of beliefs and behaviors that they use daily, without having validated the behaviors iteratively. "Do not be rude" might be a spoken rule only within the symbolic memory, and daydreamed non-symbolic scenarios could include rude actions, neither of which address the conflict with that rule.

In this model, cognitive dissonance, or inconsistent thoughts or beliefs, is rooted in the non-symbolic section. Inferring with unverified beliefs and behaviors will lead to abstracted data planes that have inconsistencies within them (One plane may contain valid and invalid beliefs.) or, more likely, there will be significant differences among the highest-level data planes or sets of the highest-level data planes. This is a challenging problem, since the highest-level data planes contain summarized or abstracted beliefs and sensed information that always match something. A tragic or otherwise significant event in one's life illustrates this mechanism. When you lose a loved one or retire from a job, for example, the belief that the loved one or job still exists is present in your daily activities. You can be taken aback when you realize that this belief is false. This belief habit can persist for many years before it is supplanted by more recent, similar copies of the beliefs and behaviors that don't include tragic or other significant information included in the top-level data planes. Behaviors and habits age slowly: They age to accommodate an aging body and cortex. You see an old person as an old person until you realize that you are older than he or she is.

Thus, I claim that I have explained, or at least accounted for, all human behavior, even though I do not, or cannot, explain why a particular reader prefers mountain climbing, reading sonnets, or interacting with other humans. There is no free will mechanism, nor does there need to be. If something is figured out through associative inferencing, there is a random component in the selection of the output that can be interpreted as free will. Choosing to watch the football game instead of mowing the lawn is not free will; the former is replete with memory-matching rewards, while the latter includes an uncomfortable energy commitment. Human behavior is either learned, mostly from the environment, or it is figured out. If a new behavior produces less-than-desired results, bad results, or great results, it can be repeated until it becomes a habit, or it can be ignored and replaced by something else.

18 SYSTEM OPERATION

The cortex is a direct current (DC) analog system that runs continuously. Data planes are continuously streamed from sensors, specialized areas, and lobes to different specialized areas and lobes. Memory is scanned against these data planes, and matches between them reinforce the data by reentering a valid data plane into memory at the next SA/TA level.

There are no switches in the cortex. At best, learned behavior, in the form of qualitative data represented in the data planes, can cause a control structure to inhibit one or more data paths to ensure the best output is used. Specifically, all of the higher-level layers of memory or specialized areas must match the input data and output data in order for an output to occur.

In early life, novel perceptions create data-specific pathways at the junctions among axons and the next downstream neurons. Data from each incoming axon is related to many of the other incoming data streams. This context development occurs at many levels and can only occur at higher levels when the lowers levels that provide data are producing valid data planes.

At each level, the data forms context connections specific to features in that data. At any level, many such specialized areas can form. Initially, spatial abstraction is mapped for the data elements from a specific sensor set. A sensor set is a set of individual sensory elements and their respective data outputs. The retina is a sensor set, with each fiber in the optic nerve being a collection of sensory element data streams.

In the next layer, the output of these spatial/temporal abstraction layers is integrated with other sensory spatial/temporal abstraction layers, further differentiating features and specialized areas. Additional data streams from the midbrain, other lobes, and output streams are integrated at another layer, as illustrated in Figure 2.

So, there are two modes of operation: early, where connections and layers are being formed, and later, where iteration validates data and inferencing integrates data from specialized areas and creates associated data.

The output data planes from spatial abstraction lack temporal context. They are not related to each other, except by sequence. This sequence is maintained as the data stream is stored in memory. There is no indexing. Input streams are sequentially matched against the stack of data planes in memory. A sufficient match, as seen by a control structure, fires the exalted data stage and rewrites the data plane in the same memory but at the proximal end. This will continue until the matches in the sequence become poor, indicating the end of the sequence. For good matches between the input streams and memory, the differences will be small. The differences are accumulated until the scaling requirement is met and are then also stored in the same memory after the data stage fires and conveys input to the next specialized area. This accumulation of differences is known as temporal abstraction. There only needs to be a few levels of such abstraction to account for all environmental conditions. More and different sensor sets could increase the number of needed layers.

Output is learned and stored with the input in integration layers. A good input match will produce immediate output. Regardless of this iterative stage, data moving to the next layer will also produce an output, and so on. These higher-level outputs are context-free data that start output streams (generally through data paths back to the original input and output level). These feedback pathways are the inferencing loops; one for symbolic data (generally language and other symbols meaningful in the environment) and one for non-symbolic data (generally sensed and abstracted symbolic data). Inferencing reinforces behavior when memory matches are good and creates new information when memory is inexactly matched. This new information must be validated internally or by the environment.

The environment rewards successful behavior, which sometimes happens indirectly through the midbrain. The data planes that generated the successful behavior are built up in memory, reducing the likelihood that they will be purged during sleep.

The cortex has no built-in goals. Energy management (supply and demand) is learned first by imitating midbrain behavior, then by imitating environmental behaviors, and finally from direct reward or punishment data from the midbrain, which creates validated behaviors or habits. Most human behavior is rooted in energy management: ensuring energy is available or minimizing energy use when it is not. Other goals are learned from the environment, probably through a language, but are only preserved if they match the expectations of the person initiating the procedure. Humans can create and execute a series of steps or behaviors, all leading to the accomplishment of a goal. Inferencing with high match rates is the execution mechanism for these plans and procedures. Associative inferencing creates plans and procedures that are not learned. The goal of the plan or procedure is to satisfy some condition that has been stored in memory earlier.

Input focus (longest memory scans) requires an energy commitment that is learned. Inferencing requires a much larger energy commitment. Energy requirements are learned and are then included in memory; this includes both cortical energy requirements and the energy required to interact with the environment.

There are no truth standards in the cortex. The human brain can believe anything is valid as long as the environment continues to reinforce it. Schools, churches, laws, family, and social living all tend to make environmentally validated beliefs more common within groups.

As the brain ages, its performance degrades somewhat evenly. Memory errors occur (these could be errors in memory entry, shortening of the memory strings, or, more likely, diminished energy used for drive scanning), inexact matches become more common because they require less energy, so downstream matches become less accurate, and thus results become less accurate. Some of the resolution can be regained by inferring the correct output through the use of memory data. Ensuring that abundant energy is available to the brain is also helpful.

Many of the contemporary mental disability definitions can be explained by deficits within this model. Some such disabilities are behavioral only, meaning memory content is invalid. Other disabilities are caused by early memory content or environment exposure creating incorrect spatial abstraction mappings. Still others are physiological errors in the cortex's structure or function. Knowing the category would certainly aid in addressing the deficit.

Self-awareness, or the "me habit", is sometimes considered an indication of consciousness and is considered an absolute mental state. Consciousness, using any definition, actually varies from moment to moment, day to day, and year to year. Things are added to memory, things are forgotten, new habits are formed, old habits are replaced, focus shifts from iterative sensing to inferring and back, and alertness changes based on the energy available and the need to conserve energy. Memory scanning is not part of consciousness. Most people are not aware of non-symbolic activity, which is why it is referred to as the subconscious. Most intelligent behavior is non-symbolic, but such results are not part of awareness until they are conveyed to the symbolic section.

Consciousness comprises a range of mental activities. An anesthesiologist can select nearly any state of consciousness for a patient, including the suspension of the "me habit." Brain injury or disease can also create deficits in mental abilities and, therefore, alter consciousness. It is noteworthy that the legal system has a very broad definition of consciousness or life, and this definition does not accommodate changes to consciousness or intelligence caused by insult to the brain beyond those caused by the grossest injuries or diseases. A blow to the head of a child could influence the future intelligence of that child, as could exposure to environmental toxins, drugs, and alcohol. The formation of spatial abstraction connections is a fast and fragile process.

19 CONSTRUCTING A COGNITIVE AUTOMATON

Human brains are constantly besieged with data about the body and its current state within the environment, as well as sensed environment data, symbolic and non-symbolic. An intelligent machine does not need most of these inputs. The goal is not to make a machine that thinks like a human, but to make a machine that can sense and store data, relate that data to other sensed and store data and infer accurate conclusions that are beyond the ability of any human cortex. It will not be encumbered by midbrain distractions and, with modest computing resources and unlimited access to past and current environmental information, it will find relationships and correlations that will solve problems that have vexed the best minds in the world. The world will see an unheralded acceleration in medical, physical, chemical, financial, and behavioral research.

With the gaps in Eugene's model filled, I constructed a method for instantiating the new model. The method does not require prior knowledge of the environment being sensed to interact intelligently within that environment. However, the environment must provide positive or negative feedback of some sort. The equivalent of energy balance management must be emulated to improve searching or inferencing results.

To differentiate my model from algorithmic and heuristic methods typically used in the current field of Artificial Intelligence, I will call it cognitive automation and the machine itself, a Cognitive Automaton (CA). Artificial intelligence requires prior knowledge of the information being processes, a CA does not.

In my model, there are only seven functions required to instantiate the intelligent mechanisms that I described above. Replicating, layering, and otherwise compounding these seven functions as environmental information that is sensed (and reinforced or discarded) will quickly establish an architecture as shown in Figure 2.

The seven functions are:

1) Sensing the environment via arrays of sensor elements, eliminating background or context via groupings of sensor data elements which consolidates and reduces array sizes into a series of data planes. The minimum required control inputs are positive and negative feedback from the environment and continuous time ticks.

2) Establish spatial and temporal context within and between data planes via data defined pathways; and accumulation and rescaling of data values transiting those pathways.

3) Perform context generation for sensor data plane input and output to the environment as well as for the integration of all other sensor and output data establishing a hierarchy of memory regions that are developed from and have pathways from lower level regions.

4) Compare memory to input and memory to output by region continuously. Memory match qualities are defined algorithmically and provide for matching across an entire region or matching at a focus within a region that satisfies match tolerance criteria. Memory match data meeting quality criteria is conveyed to the next region. Data Planes passing

through the SA/TA stage produces context dependent data which is required for this data plane matching to work.

5) Memory match criteria allows for exact matches across a region, partial matches across a region and inexact matches across a region. This will only work with context dependent data produced via SA/TA stages.

6) Exact matches within the regions will allow learned output to the environment to continue.

7) If there are no exact matches in a memory region, the match quality criteria changes automatically to permit inexact matches. This is a shift from iterative mode to inferential mode. Match quality criteria will automatically change to exact matching if memory match quality improves.

This system will learn fundamental information and learn best output, learn advanced information and advanced output; use that information to iterate (exchange) information with the environment and with sufficient memory stores, use that information to infer new information.

Data from illustrative computer programs for each function are described below.

Removing Environmental Context - Sensing

There are 120 million rods and 6 million cones in one human eye. This data is coded and multiplexed down to 1.2 million axons in the optic nerve. This shows phenomenal consolidation and encoding, considering that it is done by a biological analog system. It is important to note here that processing input data equivalent to 1.2 million fibers of all-streaming data at roughly 30 frames per second and a resolution of less than 13 bits is not particularly taxing for today's computers.

Cameras and microphones that equal eye and ear sensitivities exist today and, with a little coding or conversion of sensory outputs, equivalent data transport rates are easily achievable.

Early CA sensors will more than likely manifest as sources of digital data, such as streams of financial data, that are readily available. Experimental scientific data is another rich and readily available source of streamable data. The Large Hadron Collider facility has tremendous stores of data waiting to be analyzed and correlated, for example. Conventional methods rely on algorithms and heuristics to find such correlations. A CA would use used data directly and search for correlations based on correlations found in the past. Inferencing would allow experimental excursions from the known correlations.

Input from whatever source would simply be streams of data from the environment being sensed. A single stream of financial data is pointless, but 10,000 different streams of financial data would be ideal. The CA's greatest benefit is the integration and correlation of large sets of input data streams that are then used to generate output streams that can be validated internally or by the environment.

The easiest input to envision would be streams of video data planes—not a video signal on a cable but planes of video data in the captured sequence moving towards a spatial integration stage.

The model I show here has a modest environment but fully illustrates spatial and temporal abstraction. The environment consists of these symbols and nothing else.

Figure 5 illustrates input data. There are only three letters in this environment: A, B, and C. Each letter has four variations: Normal, Italics, Small, and Colored. Each pixel in each letter is represented by the value 100, except that, in the colored row, the pixel values in the top half of each letter are 300. This is not color in the way the human eye senses color, but it will demonstrate that shades of black will

incorporate into later stages with no loss of differentiation.

Figure 5

A stream of these symbols, in no particular order, is sensed by a sensory spatial abstractor that is different from the cortical spatial abstractor. Sensory elements are grouped or connected somewhat arbitrarily, with the best estimation being that the sensors on the periphery can have the highest grouping and sensors toward the center of the field have the least. Testing results will allow the grouping mappings to be optimized.

Figure 6 is a mapping of the sensors to the output fiber tract.

In this illustration, there are 121 sensory elements mapped to 12 input fibers or pathways. The various shaded boxes in the sensory array are 3:1 mappings to five of the 12 output fibers. The rest of the sensory array boxes are 6:1, 8:1, or 10:1 mappings to the remaining output fibers. This provides a coarse but workable resolution for this model's demonstration. When the various symbol values in Figure 5 are overlaid on the 11 x 11 sensor array, output values from the consolidation and mappings are shown for each symbol in the Results section of Figure 6.

Figure 6 — Sensor to Output Fiber Consolidation

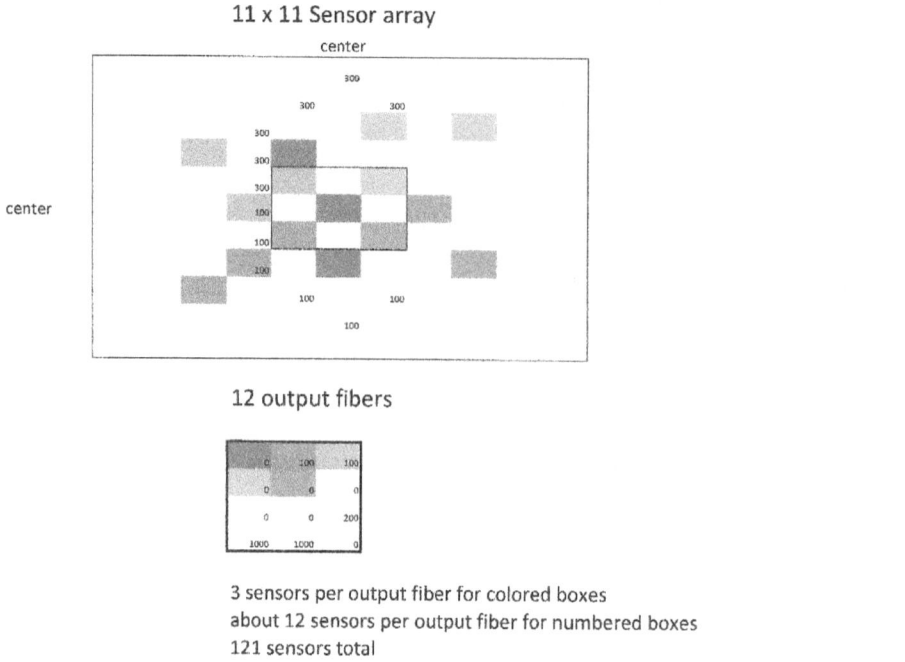

11 x 11 Sensor array

center

center

12 output fibers

3 sensors per output fiber for colored boxes
about 12 sensors per output fiber for numbered boxes
121 sensors total

Results

	Letter			Narrow			Asym			Colored		
A	0	100	0	100	200	200	0	100	0	0	100	0
	100	100	100	200	300	100	200	100	100	200	100	100
	100	0	400	100	0	200	0	0	400	100	0	400
	400	700	600	100	600	300	400	900	100	600	1200	700
B	200	100	100	200	100	100	200	0	700	400	100	300
	100	100	200	200	100	200	0	100	200	300	100	200
	100	300	200	0	100	200	300	300	400	300	900	200
	0	1000	200	0	1000	100	100	600	300	0	1800	400
C	0	100	100	0	0	100	0	200	0	0	100	100
	0	0	0	100	0	0	100	0	200	0	0	0
	0	0	200	0	100	0	100	0	0	0	0	200
	400	400	0	300	300	100	0	600	0	1000	1000	0

Re-establishing context - Spatial Abstraction

There are two separate steps in this spatial abstraction method: initial context mapping and then adjustments to the context mapping by sensed data in the data stream. Individual data element fibers that comprise a data plane pathway conduct signals from either coded sensory data or other specialized areas to a separation in the pathway. To cross this junction, each element must create one or more connections with fiber elements on the far side of the junction. Initially, the distribution is based on incrementally extending connections to nearby elements, based on the ability of these elements to accept the connection. Consider the signal as the integral part of the analog data waveform for a specific epoch that yields an energy amount. This energy amount must be dissipated among many nearby downstream fibers or elements that have a limited capacity for such dissipation. If the nearby downstream elements cannot support the dissipation needs of a specific inbound signal, the mapping process simply extends new fibers to other downstream elements that can. The initial connections are created by noise in the system or randomness in the data. The connections, in this case, can be made rapidly and show nearly a random distribution, with a bias towards geographical nearness.

Figure 7

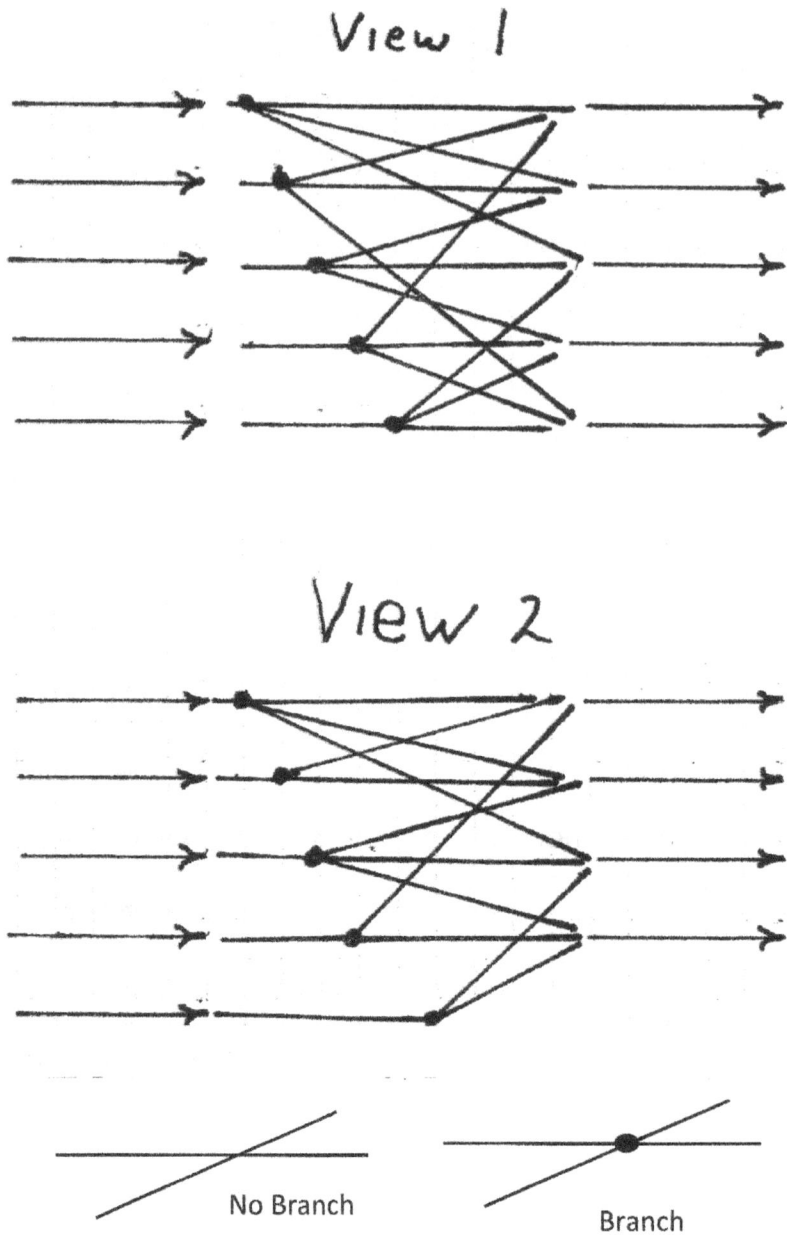

View 1

View 2

No Branch

Branch

View 1 of Figure 7 illustrates the initial random distribution from the input pathways (left) to the output pathways (right). View 2 illustrates the shift in pathway connections caused by sensed data on the input pathways. In this example, there is not enough data on the input pathways to warrant or sustain five output pathways, so the bottom connection is eventually lost. It is this connection "atrophy" that defines the edges of specialized areas in memory. The general rule is that the average amount of data elements flowing across each connection must be one-half of the scale range. This requirement may not be met on the periphery of the specialized areas.

As the sensors are exposed to uncommon symbols in the environment, more signals with higher-than-median values will appear at the junctions. Additional mappings will be created to dissipate this added energy. Eventually, the mappings will no longer be random, but will have a geographical nearness bias and will be information-driven connections.

Again, because of the way data values are redistributed, one cannot define or recognize the content of the data planes, so I cannot say that this is feature extraction.

Figure 8   Spatial Abstraction Output Values

Initial Dist	A only	A then C	Sorted Initial Dist	Sorted A only	Sorted A then C	Sorted Difference
81	32	53	13	7	13	6
33	23	38	22	10	19	9
119	66	101	28	11	21	10
99	39	63	28	13	23	10
137	73	116	29	13	23	10
100	66	104	33	14	25	11
22	18	28	35	16	28	12
60	30	50	37	16	28	12
128	63	103	42	17	28	12
53	27	43	43	17	30	13
58	13	23	46	18	32	14
96	43	72	47	18	32	14
79	38	62	47	18	32	13
54	35	60	48	19	33	13
47	34	55	49	21	38	17
111	63	99	49	21	39	18
98	64	97	51	21	43	23
46	14	56	53	23	44	21
88	64	100	53	23	44	21
96	66	103	53	24	45	21
118	71	112	54	25	45	20
114	54	85	57	27	48	21
106	58	88	57	27	50	23
146	68	107	58	27	52	25
76	27	73	58	27	53	26
123	81	98	58	27	53	26
85	16	28	58	28	53	24
90	51	83	59	29	54	25
78	51	107	60	30	55	25
53	30	48	60	30	56	26
68	55	83	61	31	57	26
112	86	110	62	32	58	27
70	48	75	63	32	60	28
113	65	103	64	32	60	28
73	33	54	66	33	62	29
127	57	88	68	33	62	28
72	38	62	69	34	62	28
62	53	79	70	34	62	28
57	29	44	70	35	63	28
78	19	63	70	37	63	27

Figure 8 illustrates example values on the first 33 fiber tracts, of a total of 120, on the output side of the spatial abstraction junctions before and after exposure to A's and C's. The initial distribution values are random values (far left column) from noise input. After repeatedly only sensing A, the distribution connections change and produce the values shown in the "A only" column. After repeatedly sensing A's and C's, the distribution output values become those shown in the "A then C" column. There is little tapering off at the ends of the columns, indicating a clear data-driven demarcation between the represented area and adjacent areas. The noticeably large difference between these two columns is an artifact of the sensor. The C input matrix happens to cover more of the broad area sensors than the A input, resulting in larger values being conveyed on the 12 sensor fibers. Further to the right, the three columns are sorted by values and a differences column is added using the full 120 output fiber tracts (i.e. the numbers in the left-hand columns will not necessarily appear in the sorted columns). The difference between the random distribution and the learned or trained distribution is apparent. The consistency of the data distribution is also evident.

Re-establishing context - Temporal Abstraction

Temporal abstraction uses the same set of fibers and signal distribution. Figure 9 is an illustration of the abstraction of 3 sequential data planes containing random values. When they are passed to the EDS the individual values transit the same pathways that define Spatial Abstraction. In this example, however, each value is decreased by the distance distribution factor in the left most column and those values are adjusted equally by a fractional multiplier to further reduce individual values so that summing only three input planes will approximately maintain the scale range of 0 to 5000. In a production system, this is not necessary. The distribution of P1 plane values to the EDS is shown in P1 Loss column (the P1 RandDist column is the random distribution of the input data values. For example, the first value in Input plane P1 (3318) appears in 3 places in the P1 RandDist EDS). The P1 Loss column shows the P1 RandDist values after the Adj. Factor has been applied. This is repeated for input planes P2 and P3. The values of the Loss planes are summed in the Sum at Firing column. As some values in this column approach the upper limit of the scale range, the stage fires, committing the Sum at Firing data to memory and starting over. The three input planes have been abstracted to the larger post-ESD plane. A gain control mechanism is necessary to ensure the abstraction levels (epoch) are high. If the sums of the input planes increase rapidly, the fractional multiplier is increased to lower the input values equally across the EDS. One method for this gain control is to sum the EDS after each input and decrease the fractional multiplier as an inverse of the change in EDS sums. If input is not unique, gain control will increase the fractional multiplier to increase the epoch. If input is unique, gain control will decrease the fractional multiplier to create more data planes for that input. The gain control variable is set by the EDS as memory is compared to input. An operational rule would be: if the input data values are rising rapidly, decrease the gain until input data starts matching memory data. A fixed upper limit will eventually fire the EDS. This feedback mechanism ensures future exposures to the same data will match the memory data. This is critical.

Figure 9 — SA/TA mapping and output example

This is a 9 input to 17 output mapping

Input Data Planes

P1	P2	P3
1131	4306	3078
4633	2607	2641
3253	2093	2252
3113	2955	4867
4378	4031	3083
1942	1764	1644
2672	228	2469
117	488	93
4305	1612	1240

Mapping distribution — each cell shows weight and Dist (From each input fiber → these are the output contributions). Column K1 is a distance distribution, L1 has a random bias applied to K1, etc.

InputPathway	F1 Dist	F2 Dist	F3 Dist	F4 Dist	F5 Dist	F6 Dist	F7 Dist	F8 Dist	F9 Dist	Output Pathway
Fiber 1	1.00 0.4706	0.87 0.4612	0.76 0.0000	0.66 0.7722	0.58 0.1355	0.50 0.2656	0.44 0.1542	0.38 0.1567	0.33 0.2926	1
	0.93 0.0549	0.93 0.5588	0.81 0.0478	0.71 0.6249	0.62 0.3992	0.54 0.2213	0.47 0.0826	0.41 0.2399	0.36 0.3344	2
Fiber 2	0.87 0.1025	1.00 0.2353	0.87 0.4612	0.76 0.5803	0.66 0.6221	0.58 0.1016	0.50 0.0000	0.44 0.0511	0.38 0.3359	3
	0.81 0.4783	0.93 0.8941	0.93 0.9333	0.81 0.4304	0.71 0.6666	0.62 0.3992	0.54 0.3161	0.47 0.0000	0.41 0.1679	4
Fiber 3	0.76 0.3571	0.87 0.3075	1.00 0.9412	0.87 0.7588	0.76 0.2588	0.66 0.6221	0.58 0.5420	0.50 0.4131	0.44 0.1542	5
	0.71 0.4583	0.81 0.1913	0.93 0.3843	0.93 0.3294	0.81 0.7686	0.58 0.3587	0.54 0.3500	0.47 0.2213	0.47 0.3305	6
Fiber 4	0.66 0.2722	0.76 0.3571	0.87 0.3587	1.00 0.0000	0.87 0.8824	0.66 0.0893	0.58 0.2903	0.54 0.1355	0.50 0.1180	7
	0.62 0.3629	0.71 0.4999	0.81 0.8130	0.93 0.8824	0.93 0.8199	0.81 0.0478	0.62 0.3750	0.54 0.2903	0.54 0.0632	8
Fiber 5	0.58 0.1016	0.66 0.1944	0.76 0.4017	0.87 0.8199	1.00 0.8824	0.87 0.3750	0.66 0.5696	0.58 0.0778	0.58 0.0339	9
	0.54 0.1897	0.62 0.1815	0.71 0.1666	0.81 0.7686	0.93 0.7686	0.93 0.3587	0.66 0.6996	0.62 0.0046	0.62 0.4355	10
Fiber 6	0.50 0.0590	0.58 0.3049	0.66 0.0389	0.76 0.8130	0.93 0.9933	1.00 0.9333	0.71 0.5124	0.66 0.6217	0.66 0.2722	11
	0.47 0.2313	0.54 0.2845	0.62 0.1089	0.71 0.6217	0.87 0.7686	0.87 0.5294	0.76 0.3826	0.71 0.6217	0.71 0.7082	12
Fiber 7	0.44 0.4078	0.50 0.0000	0.58 0.1694	0.66 0.4910	0.81 0.1025	0.93 0.8235	1.00 0.0000	0.76 0.6217	0.76 0.7588	13
	0.41 0.2570	0.47 0.1652	0.54 0.4110	0.62 0.7142	0.76 0.0833	0.87 0.5124	0.95 0.8235	0.81 0.8784	0.81 0.1435	14
Fiber 8	0.38 0.0224	0.44 0.2570	0.50 0.0885	0.58 0.6221	0.71 0.0833	0.81 0.7174	0.93 0.8235	1.00 0.0000	0.87 0.3205	15
	0.36 0.1045	0.41 0.1199	0.47 0.1377	0.54 0.3794	0.62 0.1089	0.76 0.4166	0.93 0.5739	0.93 0.2196	0.93 0.3843	16
Fiber 9	0.33 0.2731	0.38 0.0224	0.44 0.0771	0.50 0.0295	0.58 0.0677	0.66 0.2333	0.76 0.6696	0.87 0.4612	1.00 0.1176	17
	10.36 0.38									

Untrained mapping, there is no distribution bias caused by the data values. Mapping distribution is Random with a distance bias. Column K1 is a distance distribution, L1 has a random bias applied to K1, etc.

These are the output contribution: / From each input fiber:

Output Fiber	F1 Dist	F2 Dist	F3 Dist	F4 Dist	F5 Dist	F6 Dist	F7 Dist	F8 Dist	F9 Dist	Abstracted output
1	688.55	845.797	534.6	739.27	882.7	414.26	328.08	61.958	627.98	5123
2	560.6	897.995	545.37	910.55	964	406.98	310.4	62.731	633.16	5292
3	575.24	786.141	638.46	889.22	1033	387.31	290	60.978	633.35	5293
4	690.91	854.496	744.79	816.29	1046	436.23	368.06	60.5	612.52	5630
5	653.61	805.198	746.55	606.8	1075	472.88	423.81	64.342	610.82	5459
6	684.76	774.523	621.15	767.12	899.9	384.3	397.53	62.558	632.68	5224
7	627.48	818.31	615.38	1036.2	1094	385.28	376.4	61.76	606.34	5621
8	655.41	856.034	717.7	767.12	1078	378.46	382.58	63.2	599.54	5498
9	574.98	775.346	625.07	980.89	1113	429.57	455.31	63.223	595.9	5611
10	602.09	731.923	572.13	1002.5	1078	524.04	384.47	61.213	645.7	5644
11	561.86	804.511	543.36	845.77	872.5	457.64	416.52	60.915	625.45	5189
12	645.42	799.144	559.12	627.08	1033	505.99	493.33	66.282	679.52	5408
13	614.9	724	572.74	758.2	1061	370.6	493.33	66.695	685.8	5347
14	669.23	767.639	627.15	836.42	866.6	425.64	493.33	68.669	609.49	5364
15	550.59	791.883	554.53	722.2	1033	407.29	467.12	60.5	604.41	5191
16	575.86	755.679	565.61	791.44	874.5	439.09	431.7	62.542	639.35	5136
17	627.75	729.913	551.97	621.16	861.8	408.96	455.31	64.789	606.29	4928

The contribution from each input fiber to each output fiber is shown in the columns labeled "F1 Dist" for example. It means the values of any input on F1 fiber is distributed to the 17 output fibers according to the values in that column.

Figure 10 illustrates the accumulation of the three input plane values but this time information has been included in those planes in rows 3,

4 and 5. All of the distribution factors and pathways are the same as

Figure 9. The information values caused the Sum at Firing value to

change as shown in the W/Infor column. When those values are

subtracted for the original Random Sum at Firing column (and scaled

upward by 4000 to eliminate negative values) as shown in the

Difference column, a specialized area emerges (the right most grey

values). In that column a significant drop in values is seen above and

below the grey area which are the demarcation of the specialized

area created by input information.   This model can make accurate

memory matches using only the grey regions of temporally abstracted

data. This is an important attribute of Context Dependent Computing.

Figure 10

Dist	Dist Adj Factor	Input Planes			Rand distribution has a distance bias too						W/Infor Sum at firing	Random Sum at firing	Difference
		P1	P2	P3	P1 RandDist	P1 Loss	P2 RandDis	P2 Loss	P3 RandDist	P3 Loss			
0.33	0.25	994	4772	2951	994	247	4772	1187	2951	734	2168	1584	3416
0.36	0.27	854	4865	4084	854	228	4865	1296	4084	1088	2612	1726	3114
0.38	0.29	4000	4000	4000	4000	1142	4000	1142	4000	1142	3426	2909	3483
0.41	0.31	3000	3000	3000	3000	918	3000	918	3000	918	2753	4091	5338
0.44	0.33	4918	1059	4913	4918	1612	1059	347	4913	1610	3569	2937	3368
0.47	0.35	4000	3000	2000	4000	1405	3000	1053	2000	702	3160	2690	3530
0.50	0.38	854	4865	4084	854	321	4865	1830	4084	1536	3688	2437	2749
0.54	0.40	4000	4000	4084	4000	1612	4000	1612	4084	1830	3688	4107	3270
0.58	0.43	3000	3000	3000	3000	1296	3000	1296	3000	1296	3887	5776	5889
0.62	0.46	994	4772	2951	994	460	4772	2208	2951	1365	4034	2948	2914
0.66	0.50	4764	4929	3474	4764	2362	4929	2444	3474	1722	6528	3770	1243
0.71	0.53	2755	3202	4197	2755	1463	3202	1701	4197	2229	5394	5049	3656
0.76	0.57	854	4865	4084	854	486	4865	2769	4084	2324	5579	3686	2107
0.81	0.61	3000	3000	3000	3000	1829	3000	1829	3000	1829	5488	8156	6668
0.87	0.65	854	4865	4084	854	558	4865	3178	4084	2668	6405	4232	1827
0.93	0.70	3140	3240	3301	3140	2198	3240	2268	3301	2311	6777	8509	5733
1.00	0.75	1123	106	3220	1123	842	106	80	3220	2415	3337	4595	5259
0.93	0.70	994	4772	2951	994	696	4772	3340	2951	2066	6102	4459	2357
0.87	0.65	4000	3000	2000	4000	2613	3000	1960	2000	1307	5880	5005	3125
0.81	0.61	4918	1059	4913	4918	2999	1059	646	4913	2996	6640	5465	2825
0.76	0.57	1123	106	3220	1123	639	106	60	3220	1833	2532	3487	4955
0.71	0.53	3000	3000	3000	3000	1594	3000	1594	3000	1594	4781	7105	6324
0.66	0.50	2755	3202	4197	2755	1366	3202	1587	4197	2081	5034	4713	3679
0.62	0.46	3140	3240	3301	3140	1453	3240	1499	3301	1527	4480	5625	5145
0.58	0.43	4918	1059	4913	4918	2124	1059	457	4913	2122	4703	3871	3168
0.54	0.40	4764	4929	3474	4764	1920	4929	1987	3474	1400	5307	3065	1758
0.50	0.38	4000	3000	2000	4000	1505	3000	1129	2000	752	3386	2882	3496
0.47	0.35	3140	3240	3301	3140	1103	3240	1138	3301	1159	3399	4268	4869
0.44	0.33	1123	106	3220	1123	368	106	35	3220	1055	1458	2008	4550
0.41	0.31	4764	4929	3474	4764	1457	4929	1508	3474	1063	4027	2326	2299
0.38	0.29	4918	1059	4913	4918	1404	1059	302	4913	1403	3109	2559	3450
0.36	0.27	4764	4929	3474	4764	1269	4929	1313	3474	926	3508	2026	2518

Symbol Recognition – Memory Matching

A simple memory match and store is illustrated in Figure 11. The input data plane is represented in the shaded column on the left. Memory data planes are to the right.

The input plane matches memory plane 8, which causes the stage to fire and commit the input plane to memory at the beginning of memory.

Figure 11   Memory Matching

Bottom 20 memory elements with pre-identified output values

100	26	30	13	12	15	32	17	23	26	30	19	12	15	17	17	13	13	26	26	15	17	30
101	17	30	13	9	15	28	17	11	17	30	13	2	15	9	9	13	13	17	17	15	14	22
102	20	11	8	7	6	71	15	18	20	11	8	7	6	7	7	8	8	20	20	6	15	11
103	20	20	11	7	10	20	13	20	20	20	11	7	10	7	7	11	11	20	20	10	13	20
104	28	23	17	9	17	30	10	28	28	23	17	0	13	9	9	17	17	28	28	13	8	14
105	28	19	14	8	9	22	13	23	28	19	14	8	9	8	8	14	14	28	28	9	10	11
106	15	2	3	6	2	13	18	10	15	2	3	6	2	6	6	3	3	15	15	2	18	2
107	33	28	16	15	13	43	25	33	33	28	16	15	13	15	15	16	16	33	33	13	25	28
108	22	30	9	11	15	26	20	18	27	30	9	11	15	11	11	9	9	22	22	15	30	30
109	14	21	4	8	11	18	18	11	14	21	4	8	11	8	8	4	4	14	14	11	18	21
110	25	10	9	9	5	24	13	18	25	10	9	9	5	9	9	9	9	25	25	5	13	10
111	28	34	13	13	14	33	18	23	28	34	13	13	14	13	13	13	13	28	28	14	20	43
112	35	22	14	8	12	23	11	32	35	22	14	8	17	8	8	14	14	35	35	12	11	22
113	33	30	16	13	15	38	24	30	33	30	16	13	15	13	13	16	16	33	33	15	27	38
114	28	13	9	7	8	19	11	20	28	13	9	7	8	7	7	9	9	28	28	8	11	13
115	28	35	11	15	15	37	24	23	28	35	11	15	15	15	15	11	11	28	28	15	27	43
116	30	27	22	9	12	32	12	31	33	27	22	9	12	9	9	22	22	33	33	12	9	18
117	19	3	6	7	3	28	23	15	19	3	6	7	3	7	7	6	6	19	19	3	21	3
118	17	8	10	5	3	28	22	13	17	8	10	5	3	5	5	10	10	17	17	3	12	8
119	21	33	12	13	13	18	14	21	21	33	12	13	13	13	13	12	12	21	21	13	14	33
120	20	23	9	8	19	19	11	10	20	23	9	8	13	8	8	9	9	20	20	13	11	23
Output >	B1	C4	C3	C2	C1	B4	B3	B2	B1	A4	A3	A2	A1	SeqC2	SeqC3	SeqA3	SeqC3	SeqB1	SeqB1	SeqA1	B3	C4

Partial matching using standard deviation as a matching algorithm is shown in Figure 12.

Figure 12 — Partial Match Illustration

Memory Position	Input (A1)	1 SeqC2	2 SeqC2	3 SeqA3	4 SeqC3	5 SeqB1	6 SeqB1	7 SeqA1	8 B3	9 C4	10 C2	11 C2	12 B1	13 C1	14 B4	15 B2	16 C3	17 A4	18 A2	19 A3	20 A3	21 A1
100	30	12	18	28	3	13	17	26	4	30	0	18	26	4	15	-18	13	40	-10	3	28	0
101	24	7	18	17	8	13	8	17	14	11	2	18	18	12	25	11	13	17	-4	4	17	0
102	16	7	9	17	8	11	8	17	15	11	2	9	17	13	21	11	13	28	-2	-7	17	0
103	22	7	15	18	2	11	8	20	-3	20	2	15	20	10	20	11	8	11	-8	2	20	0
104	16	7	15	18	-2	11	4	19	-19	19	3	15	19	12	23	18	8	22	-6	2	-18	0
105	18	6	13	11	8	9	5	2	18	20	21	13	13	6	14	15	9	11	-6	1	6	0
106	23	6	17	6	17	3	8	15	-18	33	6	17	13	11	23	9	10	9	-16	6	17	0
107	36	15	21	35	16	20	33	33	25	28	8	21	21	13	43	10	16	52	-16	37	-1	0
108	32	11	21	24	8	22	10	22	30	21	6	21	22	15	19	7	14	33	-9	24	1	0
109	27	8	19	19	8	23	14	27	18	32	8	19	33	11	18	13	9	41	-20	41	8	0
110	10	9	1	9	-1	23	-15	25	10	10	9	19	14	15	18	13	9	23	4	11	35	0
111	34	16	18	35	-1	16	36	-2	11	24	8	16	18	18	23	-14	18	47	-11	9	32	0
112	24	8	16	23	2	14	34	-11	43	22	-8	16	16	11	12	32	-7	19	-7	4	23	0
113	42	16	26	41	21	21	41	35	24	38	16	26	35	12	41	-7	14	60	-14	4	35	0
114	15	7	8	16	1	26	-12	28	13	13	2	8	28	8	19	20	9	19	18	1	42	0
115	33	7	16	32	9	26	28	33	43	43	-10	16	41	5	39	-23	11	21	-13	9	41	0
116	16	7	9	19	-7	9	-8	24	13	24	-8	9	23	3	15	18	17	36	24	14	23	0
117	30	7	23	14	-1	24	-16	19	-2	23	8	23	33	8	52	1	16	15	-13	14	32	0
118	12	5	7	10	-5	19	11	30	7	18	3	7	19	5	26	-7	24	32	14	16	16	0
119	20	13	8	25	2	11	-5	11	-5	9	-13	8	21	3	28	-7	12	30	-13	10	25	0
120	20	8	12	23	9	20	-1	20	11	23	-3	12	20	13	19	21	9	27	15	12	30	0
Standard Deviation		6.8	6.8	5.1	7.5	6.8	6.8	0	4.7	8.2	6.8	6.8	6.8	6.4	6.1	6.8	7.5	4.5	5	5.1	5.1	0

Again, only the bottom 20 rows of input fibers and memory are shown. The grey columns are the differences between the input value (the column labeled "Input") and the memory plane to its left. The number at the bottom of the grey column is the standard deviation of the differences. Memory plane 7 is an exact match. Any standard deviation below 5.1 also matches the input to some degree. These columns encode data for variations of A. The input is variation 1, column 3 is variation 3, column 17 is variation 4, and so on.

Interestingly, column 8 produces a standard deviation of less than 5.1, so it is also a match, but the data in this memory plane encodes variation 3 of B. This illustrates an imperfect or associative match. No other variation of B or C matches the input.

This example also illustrates partial matches, in that all of the A versions in memory are matches to input version 1 of A.

If I use all of the 120 fibers and memory elements in the computations, B3 does not appear as an inexact match in this example. Improving the resolution of the input sensors will also narrow the range of inexact matches.

Using a standard deviation threshold works well for this example and is similar to how the cortex performs memory matching. The cortex and my model use other variations of regional and specialized area memory matches that permit loci with strong matches to influence or start a cascade of exalted data stage firings. This method allows for inhibiting input from other areas to prevent or delay the firings. Using an incorrect or incomplete memory-matching algorithm in a computer model results in either catatonic behavior or non-useful inferencing, both of which are easily recognized. Corruption of memory competency is harder to detect in learning systems and a little easier in more mature systems with functioning output. Any adjustments of the memory-matching algorithm require restarting the learning process with a known competent memory.

Iteration and Inferencing

These illustrations show that associative thinking, as postulated by Eugene, is non-symbolic inferencing that allows for imperfect matches.

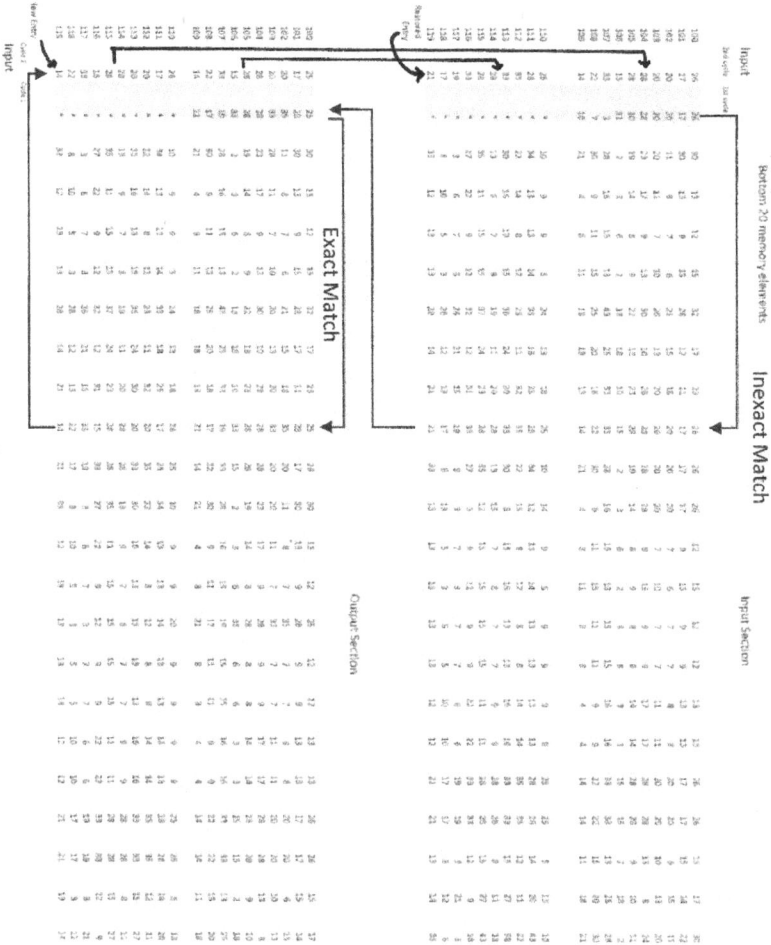

Figure 13   Inferencing to restore faded memory

Figure 13 illustrates inferencing with partial matches to restore faded memory. The top 20 rows are part of the input section of memory and the bottom 20 rows are part of the output section. Each of these sections has input columns and output columns (shown in grey). Fiber mapping between sections does not have to be exact. In the input section, the input data for the first cycle is shown in that column. It inexactly matches with memory column 8. The associated output data in that same data plane is conveyed to the input side of the output section, where it exactly matches column 8 of that section. The associated output data plane is conveyed to the input side of the input section or second cycle where, along with the original (grey portion) of the output, it is entered into memory. That memory entry exactly matches memory column 8, where cycle 1 only matches part of it.

The data flow for such inferencing is shown above in Figure 2. Such an architecture allows for internal iteration and inferencing with exact and inexact matches, including the procedural variation that I described in Chapter 10.

With such an imperfect matching capability, associative inferencing can occur. The output from a specialized area can imperfectly match a data plane in the next specialized area, generating an output that is different, but possibly still valid, for the next inference. The inference thread wanders through memory regions until an output iteratively match the input stream that started the thread or other input data becomes prevalent and inferencing stops.

If there is an explicit goal, it is very likely staged in symbolic memory or in the environment as a visual, audible, or physical goal. Any inferential output is tested in memory through further inferencing, or it is tested in the environment and is either reinforced or not.

Figure 14 is a data flow diagram for a minimal system.

Figure 14 Data Plane Flow Diagram Example

	Region 1 Input Memory			Region 2 Integration Memory		Region 3 Output Memory				
Input 1	Input 2	Layer 2	Layer 3		Layer2	Output 2	Output 1	DP11	DP11	DP11
800	132	8064	21	3 17798232	8232	818	201	959		
321	689	8866	56	46 68668703	8703	638	245	626	241	
790	350	2069	66	23 20696121	6121	268	772	441	247	
649	182	1557	85	11 15577147	7147	825	320	626		
991	513	2486	35	60 24868714	8714	236	630	463	606	
201	920	2575	15	54 17794798	4798	157	121	676	183	
905	385	7123	52	90 71234499	4499	610	370	696	385	
783	404	1044	89	87 10447066	7066	896	977	695	348	
105	868	5115	36	94 51159777	9777	602	738	230	783	
112	134	6674	24	98 66742965	4275	748	353	981	382	
334	332	1779	44	99 17792965	2965	346	509	371	139	

EDS

SA/TA 000 000 0000 00 00000000 0000 0000 196 491

SA/TA 000 000 SA/TA SA/TA SA/TA

Environment or System Reinforcement (1 to 100)

		610	370	799	155	155
		896	977	558	156	978
		602	738			
		748	353			

Sensor 1   Sensor 2

Sensor 1	Sensor 2
76	27
76	74
73	5
67	55

Sensor 1 Sensor 2 Inverted reinforcement values(100-value)
(Values shown contain no information, they are only symbols that can match or not match an input)

From left to right there are two input sections for sensors, and integration section for those inputs, an input/output integration section, an output integration section for the two output sections and the two output data sections.   The row of zeros represents the

Exalted Data Stage. Data rows above them are in memory. Data rows below them are inbound and outbound data. The additional two layers behind the exposed layer represent part of the data planes for each single data value on the exposed layer. The DP11 labels above the numbers with the grey background indicate two additional values that make up that data plane. The ellipsis indicates the rest of the elements that make up DP11, continue into the page. Individual data planes in this representation are not two-dimensional matrices. They are one dimensional and project into the page. (201, 959 and 241 are the first three entries in the right-hand topmost output data plane.) The actual topology of the data planes must minimize connection fiber length in a true analog system. In this model, that is not required.

Each input and output section contains two columns of numbers. The numbers above SA/TA and exalted data stage zeros are memory entries from that SA/TA stage. The input and output integration sections and the overall integration section uses the same representation but with wider data planes. Data flow is always: sensors to SA/TA, to Input integration Layer 2, to integration without SA/TA Layer 3; and output data from memory to SA/TA for output integration, to integration without SA/TA. Output data is consolidated just like input data and associated with the input data via the integration layers. Output data is not switched. Like input data, it branches and flows to two places.

The environment must provide feedback to the system for all outputs made by the system. The feedback is generally sensed data and can also be symbolic data, or input from the midbrain (pain or pleasure or the many variations of midbrain output to the cortex). The sensed feedback is consolidated, integrated (not shown), and stored in the integration section Layer 3 memory alongside the integration data pair that generated that feedback. Those values are also inverted, in this example Maximum Scale Range minus the value equals the inverted value, and stored in the same sensor integration data plane that generated the match in the integration layer. Obviously, there are time lags between output and feedback. Temporal realignment is accommodated by Temporal abstraction, buffering, or pseudo-pausing in the Integration layers done by generating more copies of the same data planes.

In operation, sensor data is presented at the EDS and compared to memory; a match will produce an associated output when the EDS fires or commits data to memory. A lower quality match in any section will cause inferencing or iteration to begin. A memory match control system determines when inferencing or iteration begins. A higher quality match will simply allow output to continue for the current input.

Figure 14 is illustrative. The number of sections is limited only by input and output requirements and implementation hardware. Data plane widths, scale ranges, and memory depths are only limited by the same things.

Memory Control System

There are several rules the system must follow:

During memory scanning, each data plane in a region is presented to the ESD in the order it was stored. Memory reading and writing are both sequential, i.e. Memory is not addressable.

Memory scanning and control is independent of ESD firing.

Memory scans continuously, it cannot stop. Think of the three memory regions in this example as three wheels on a slot machine. Each wheel has data planes inscribed in rows around the wheel. All wheels are spinning continuously. The left wheel matches input to memory and sends the differences to the center wheel. The output wheel sends data to the output mechanisms and the center wheel.

The center wheel spins and fires the ESD when the best matches are made for both the input and output on the center wheel. The three wheels are thereby synchronized by virtue of their current position when the ESD fires. They stay synchronized until a mismatch occurs, more than likely with environmental feedback data or a change in sensed data.

Iteration and inferencing are effective only by desynchronizing memory region scanning and by data section layouts so that there can be foci for regional matching in several specific regions.

The control system is simple. Data differences from the comparators are sensed and tested against a threshold. The threshold can vary based on experience, energy expenditure, or epoch. The best matches will occur in a small area of the data plane, such as a specific specialized area. Just sensing a percentage match or average differences match across the entire data plane will not work.

The control mechanism simply senses the output of some or all comparators and, when a set of comparators in close proximity, match within some tolerance, the regions' data stage is fired. Tolerance is tight until matching diminishes. Tolerances are loosened to increase the match rate and initiate inferencing.

If a strong match occurs in regions 1 and 2, the ESD doesn't need to fire. The input and output have been experienced before and the output is valid. The output will continue to stream as long as there is a strong match at the input and integration regions. The output sections do not require a memory match (there is nothing to match to.) (Trial outputs are stored in memory during an ESD firing.) Output data flows toward the output mechanisms and to the SA/TA pathways for abstraction and further integration.

An EDS memory match across all sections, except output, will cause output from output memory simply because it is in the same data plane and all sections are scanning memory in synch.

The input sections and the input integration section data planes always scan together as they are the same region. The output sections and the output integration section data planes always scan together as a region. The Layer 3 integration layer synchronizes its scanning with input the other two regions. There will always be something that matches or almost matches input integration data and that integration layer match will have integration output associated with it. Feedback from the environment is needed to change that output association as illustrated in Figure 15.

If, for a given input, the memory match in the integration section is poor, inferencing will begin automatically. This is done by reducing the memory matching requirements in the integration section gradually; a little for each complete memory scan. The actual amount will vary by implementation. An inexact input match will fire the EDS for the all integration sections, as they are still scanning in synch. If an ESD wide match is not made, the integration layer is desynchronized and the process repeats until a sufficient match is made. This is inferencing. If a procedure is being executed, the Context Free Data portion of the ESD data matches must be exact.

Developing and operating a CA computer program

Sets of sensors that produce output on parallel fibers, define the layer 1 of the computer data flow model. Cameras and sound sensors are two examples. Any stream of related data represented in matrices will also serve as input. The flow of input data is continuous; memory scanning is continuous.

Each stream of input matrices is conveyed to an SA/TA stage. SA/TA output data are stored in memory in the sequence they are conveyed to that stage. Output of those stages are conveyed to a next layer integration set of SA and TA stages. (Input and Output integration regions (Layer 3) do not need the extra SA/TA stage but do need the mechanism that provides the differences between the input and memory that are then conveyed to SA/TA mechanisms in other regions.) Other environmental data such as reward and punishment, environmental time, and energy expenditure, if applicable are also conveyed to the initial and integration SA/TA stages and regions. Output is a stream of matrices that provide visual, audible or mechanical results to the users. Earlier illustrations include examples of input and output matrices.

Figure 15 is an example of iteration.

The memory columns contains sensor, integration and output memory. Assume the output values are valid and the input values result from the exposure of a sensor to an environment. (These may be the representation of utterances of sound, for example) The integration layer is a combination of input and output memory entries assembled from random associations that eventually become valid associations using the iterative process.

In this illustration, "slo" is presented at the sensor integration memory. It is not found in Sensor memory but is an inexact match to "sln" in sensor and integration memory. "sln" is associated with output 789.

That trial output is made and results in a 9 feedback from the environment which when inverted is 91. 9 is stored with "slo" in the integration region while 91 is stored with "slo" in the sensor integration region. (Again, the timing is easily accomondated with the mechanisms I have already described). In my model, the integration memory is ordered by reinforcement value with the lowest reinforcement value at the distal end of memory. This improves the likely hood of a better but inexact match being made before the lower reinforcement value associated with the exact match to input is encountered.

The next encounter with a sensed "slo" (Figure 15B) results in 91 "slo" presented to the integration memory where it is a partial match to 67 "slm" (reading from the bottom up) which has an associated output of 456. That output is tried and the result is a reinforcement value of 80 for integration and 20 for sensor integration. 80 "slo" 456 is a valid association. The inverted reinforcement value provides no significant information in this step. It just insures that a partial match of "slo" to anything else is weighted or biased differently.

The next encounter with a sensed "slo" (Figure 15C) results in 20 "slo" being presented at the integration ESD where the exact match with "slo" results in the previously proven output of 456.

A new valid output has been associated with a nearly novel input.

Exact matches cause the ESD to fire. In lieu of exact matches, partial matches across wider areas of the ESD will cause it to fire.

Interestingly enough, in this model, output is reusable so sets of valid outputs or output streams can be reused for different but somehow similar inputs. The new associations are made in integration memory and not output integration memory.

Figure 15A

Associating Input and Output Example

Input Memory			Integration Memory		Output Memory
	Sensor	Environm ental			
Integrated	Reinforce	Reinforce			
Sensor	ment	ment		Associate	
Memory	Value	Value	Input	d Output	
			12 slm	122	959
sln			22 kk	456	456
skk			46 sln	789	789
			67 slm	456	111
			89 cc	111	122

ESD   slo

slo is an inexact match to sln and is associated with 789 output
(this match is illustrative only)

			9 slo	789	
			12 slm	122	959
sln			22 kk	456	456
skk			46 sln	789	789
slo	91		67 slm	456	111
			89 cc	111	122

ESD   slo

Output is tried and results in 9 reinforcement values from environment
Integration Memory region ESD fires storing slo 789 association with 9 Reinforcement value
Reinforcement value determines the location in memory (lower values at distal end)
Input Memory region ESD fires storing inverse of 9 with slo

Figure 15B

Next encounter with slo

		9 slo	789	
		12 slm	122	959
sln		22 kk	456	456
skk		46 sln	789	789
slo	91	67 slm	456	111
		89 cc	111	122
ESD slo		91 slo		

The next encounter with sensed slo results in 91 slo presented at the Integration Memory ESD
which is a best but inexact match to both 67 and slm with a resulting output of 456
scanning is from the bottom up and both reinforment and sensor values must match within
matching limits

		9 slo	789	
sln		12 slm	122	959
skk		22 kk	456	456
slo	91	46 sln	789	789
slo	20	67 slm	456	111
		89 cc	111	122
ESD slo		80 slo	456	

Output is tried and results in an 80 reinforcement value from the environment
Integration Memory Region ESD fires storing slo 456 with a reinforcement value of 80
Reinforcement value determines the location in memory (Higher value at ESD end)
Input Memory region ESD fires storing inverse of 80 with slo

Figure 15C

Next encounter with slo

			9 slo	789	
sln			12 slm	122	959
skk			22 kk	456	456
slo	91		46 sln	789	789
slo	20		67 slm	456	111
			89 cc	111	122
			80 slo	456	
ESD	slo		20 slo		

The next encounter with sensed slo results in 20 slo presented at the Integration Memory ESD
Because slo is an exact match with the first slo encountered, a regional match fires the ESD with
a 456 output which is an experienced valid output association.

Overtime, reinforcement values for a given input will converge on the middle value of the range
Learned valid outputs are reusable, (i.e. can be associated with other sensed input)
Reinforcement from the environment can be an operator providing positive or negative
feedback or sensed data showing that the output is diverging from prior experience
(in sensor memory) with that output or inferential reinforcement feedback from other regions

20 EXCEEDING HUMAN INTELLIGENCE

Unencumbered by physiological constraints and the distraction of

sustainment and wellbeing, a thinking machine can focus on learning

what the world already knows, correlating that information and data,

and reaching new useful conclusions. One important conclusion that

will frequently be made is that not enough data exists to figure something out. In the context of all known things, this will bring better direction to future studies, experiments, and research. With this approach, machine correlations and associations will diminish any such furtive attempts by humans.

Machines can now be made that outperform humans by increasing scale range, memory depth, sensor types, sensor resolutions, specialized area sizes, and temporal integration epochs or by rearranging the various processing layers. Today's computers are more than capable of executing this method at these higher capacities.

Imagine a machine that has read and understood the validity of every book ever written, understands all languages, and has access to all data from all of the sources that man has created. Such a machine could iterate with humans to validate inferential conclusions or gaps in data, and then contemplate (non-symbolic iterating) and realize things that would take humans centuries to figure out.

Such implementations are realistic and close at hand.

Early automatons, without language, can directly analyze the results of scientific experiments such as the Large Hadron Collider, which is currently producing more data than can be evaluated by scientists and their automated tools within the next five or 10 years. By being taught the conclusions that man has learned from early data, CAs can learn to "see" correlations in later data. This is attractive because learning can be fast and nearly automatic.

Learning a language or human-like behaviors will initially be an arduous task. Each layer of data planes would require specific feedback as trial outputs are generated, and humans would have to provide this feedback. Changing sensor coding or adjusting internal structures would likely require starting over.

After learning a language, a CA might study human diseases and develop strategies, or even treatments, for correcting the underlying causes.

Human replacement CAs could be developed to perform all tasks that are currently being performed by humans. The cost of space exploration would drop significantly because humans would not travel; only better-performing CAs would make the trips. These machines would not need limited acceleration or mission duration or life support, nor would they need to return home.

This level of machine intelligence is within reach in the very near future.

21 CONCLUSION AND GOVERNANCE

The mathematics presented here are sound as is the concept of context dependent data processing. Some of these constructs are partially based on my father's original ideas and the physiology of the human cortex. I claim that my model can account for all intelligent human behavior. Working backwards, some interesting ideas and speculation about evolution, behavior and the ever-shifting human genetics emerged.

I made 10 copies of the computer program that produced the spatial and temporal abstraction results shown above. The copies were initialized identically and were then exposed to the same ABC data as the example above. Each developed a set of unique mappings from the sensor to memory. When exposed to the same set of individual letters after the abstraction pathways were established, the results of each program were different! The standard deviations of the differences between the sensed input planes and memory varied slightly and with the same threshold, one copy did not make the inexact match with "B." Two copies did not recognize the italic "C." While most of this variation is due to the coarseness of the sensor count and mapping, it does demonstrate the innate indeterminacy of cognitive automata. It also demonstrates that partial and inexact matching using context dependent data enables associative or inferential thinking.

In current CA's, indeterminacy does not matter because the developer knows everything that the CA is exposed to. When symbolic iteration begins i.e. the system has a language and can read, developers and trainers will have a difficult time knowing everything that is in the systems memory. When non-symbolic inferencing begins, the developers and trainers will only know what is in memory when and if the CA tells them.

While the tremendous benefit of such super thinking machines will certainly change the world, there must be careful construction and supervision to ensure instantiations generally focus on things that humans cannot do but which, once done, will supplement human endeavors and not replace them. Capabilities in excess of these must remain unused or, at least, in check.

To that end, any unlicensed use of the intellectual property herein is prohibited by its owner. OlneyLabs, LLC of Sammamish, Washington is the only authorized licensing agent. The intellectual property herein

is protected by the patent, copyright, and trademark laws of the
United States of America and other nations.

There is so little time…

www.ingramcontent.com/pod-product-compliance
Lightning Source LLC
Chambersburg PA
CBHW061127210326
41518CB00034B/2541